CONTENTS

INTRODUCTION

Let's keep it simple: I want you to enjoy math. But the fun isn't in using songs to memorize formulas or doing contrived examples about ice cream cones. It's in the exhilaration of making an idea click for yourself.

The goal is to communicate like a real person, share insights, and be mercifully succinct. Let's bring on the "aha!" moments!

How To Use This Book

This isn't a reference or workbook. It's about getting ideas to click; I want concepts like e and i to be as natural to you as "a circle is round". More importantly, I want to spread the notion that any subject *can* be understood intuitively if we find the right approach.

Here are a few practical ways to use the book:

- Entertainment: Yes, really. Math is fun when you see how ideas fit together and evolve. Did you know that negative numbers were only created in the 1700s, and were considered absurd? That imaginary numbers had the same fight when introduced? Me neither.

- Study Supplement: If you're a student, read this along with your textbook. Keep the analogies in your head as you do example problems to see how they fit into place.

- Teaching Aid: Teachers, parents, and other educators: feel free to incorporate the text, analogies or diagrams into your learning materials. Analogies and visualizations help enormously with puzzling concepts like imaginary numbers. These chapters incorporate the feedback of many thousands of readers.

- Learn How to Learn: The essays highlight my favorite learning method: get the context of an idea, formulate analogies, and cover examples using those analogies. This learning technique works with many subjects, not just math.

Feedback

Feedback and comments are welcome: kalid.azad@gmail.com, or on the website http://betterexplained.com.

Colophon

This book was typeset in LaTeX using the excellent memoir package and a chapter style based on daleif1. While hairy at times, LaTeX is unmatched when writing anything related to math.

The Legal Stuff

1

DEVELOPING MATH INTUITION

Our initial exposure to an idea shapes our intuition. And our intuition impacts how much we enjoy a subject. What do I mean?

Suppose we want to define a "cat":

- **Caveman definition:** A furry animal with claws, teeth, a tail, 4 legs, that purrs when happy and hisses when angry...

- **Evolutionary definition:** Mammalian descendants of a certain species (*F. catus*), sharing certain characteristics...

- **Modern definition:** You call those *definitions*? Cats are animals sharing the following DNA: ACATACATACATACAT...

The modern definition is precise, sure. But is it the *best*? Is it what you'd teach a child learning the word? Does it give better insight into the "catness" of the animal? Not really. The modern definition is useful, but *after* getting an understanding of what a cat is. It shouldn't be our starting point.

Unfortunately, math understanding seems to follow the DNA pattern. We're taught the modern, rigorous definition and not the insights that led up to it. We're left with arcane formulas (DNA) but little understanding of what the idea *is*.

Let's approach ideas from a different angle. I imagine a circle: the center is the idea you're studying, and along the outside are the facts describing it. We start in one corner, with one fact or insight, and work our way around to develop our understanding. *Cats have common physical traits* leads to *Cats have a common ancestor* leads to *A species can be identified by certain portions of DNA*. Aha! I can see how the modern definition evolved from the caveman one.

But not all starting points are equal. The right perspective makes math click — and the mathematical "cavemen" who first found an idea often had an enlightening viewpoint. Let's learn how to build our intuition.

1.1 What is a Circle?

Time for a math example: How do you define a circle?

Defining a Circle

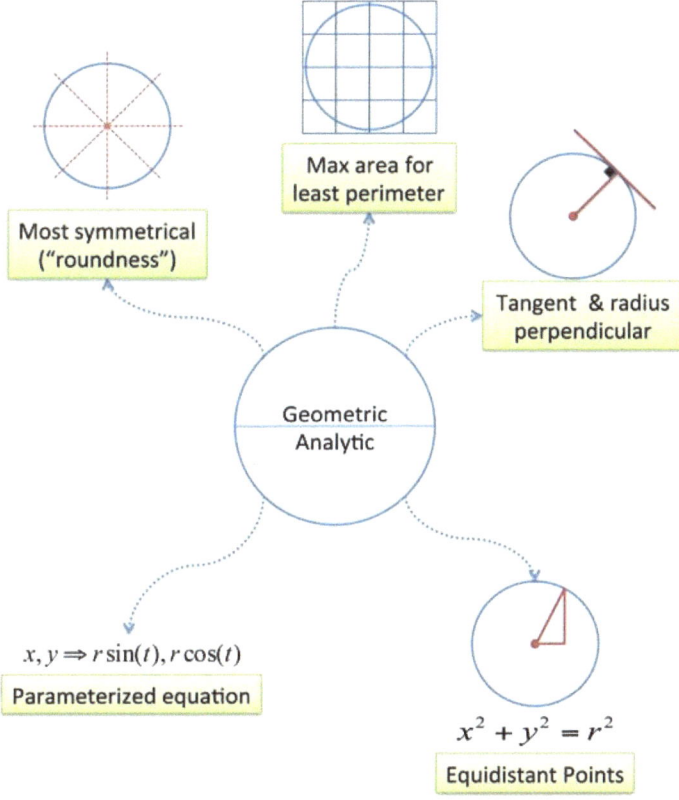

Most symmetrical ("roundness")

Max area for least perimeter

Tangent & radius perpendicular

Geometric
Analytic

$x, y \Rightarrow r\sin(t), r\cos(t)$

Parameterized equation

$$x^2 + y^2 = r^2$$

Equidistant Points

There are seemingly countless definitions. Here's a few:

- The most symmetric 2-d shape possible

- The shape that gets the most area for the least perimeter (the isoperimeter property)

- All points in a plane the same distance from a given point (drawn with a compass, or a pencil on a string)

- The points (x,y) in the equation $x^2 + y^2 = r^2$ (analytic version of the geometric definition above)

- The points in the equation $r \cdot sin(t), r \cdot cos(t)$, for all t (really analytic version)

- The shape whose tangent line is always perpendicular to the position vector (physical interpretation)

The list goes on, but here's the key: the facts all describe the same idea! It's like saying 1, one, uno, eins, "the solution to 2x + 3 = 5" or "the number of noses on your face" — just different names for the idea of unity.

But these initial descriptions are important — they shape our intuition. Because we see circles in the real world before the classroom, we understand their "roundness". No matter what fancy equation we see ($x^2 + y^2 = r^2$), we know deep inside that a circle is round. If we graphed that equation and it appeared square, or lopsided, we'd know there was a mistake.

As children, we learn the caveman definition of a circle (a really round thing), which gives us a comfortable intuition. We can see that every point on our "round thing" is the same distance from the center. $x^2 + y^2 = r^2$ is the analytic way of expressing that fact (using the Pythagorean theorem for distance). We started in one corner, with our intuition, and worked our way around to the formal definition.

Other ideas aren't so lucky. Do we instinctively see the *growth of e*, or is it an abstract definition? Do we realize the *rotation* of *i*, or is it an artificial, useless idea?

1.2   A Strategy For Developing Insight

I still have to remind myself about the deeper meaning of *e* and i — which seems as absurd as "remembering" that a circle is round or what a cat looks like! It should be the natural insight we start with.

Missing the big picture drives me crazy: math is about *ideas* — formulas are just a way to express them. Once the central concept is clear, the equations snap into place. Here's a strategy that has helped me:

- **Step 1: Find the central theme of a math concept.** This can be difficult, but try starting with its history. Where was the idea first used? What was the discoverer doing? This use may be different from our modern interpretation and application.

- **Step 2: Explain a property/fact using the theme.** Use the theme to make an analogy to the formal definition. If you're lucky, you can translate the math equation ($x^2 + y^2 = r^2$) into a plain-english statement ("All points the same distance from the center").

- **Step 3: Explore related properties using the same theme.** Once you have an analogy or interpretation that works, see if it applies to other properties. Sometimes it will, sometimes it won't (and you'll need a new insight), but you'd be surprised what you can discover.

Let's try it out.

1.3   A Real Example: Understanding *e*

Understanding the number *e* has been a major battle. *e* appears everywhere in science, and has numerous definitions, yet rarely clicks in a natural way. Let's build some insight around this idea. The following section has several equations, which are simply ways to describe ideas. Even if the equation is gibberish, there's a plain-english concept behind it. Here's a few common definitions of *e*:

The faces of *e*

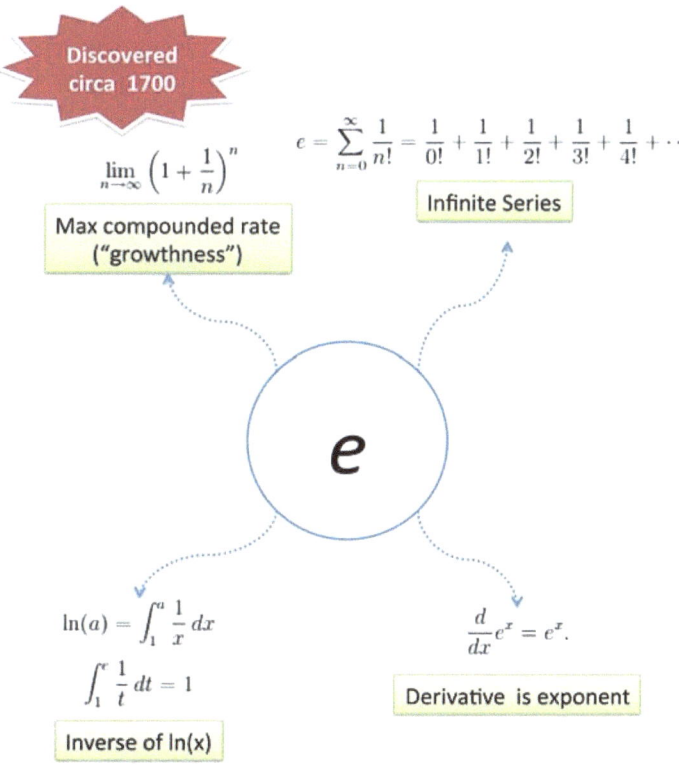

The first step is to find a theme. Looking at *e*'s history, it seems it has something to do with growth or interest rates. *e* was discovered when performing business calculations (not abstract mathematical conjectures) so "interest" (growth) is a possible theme.

Let's look at the first definition, in the upper left. The key jump, for me, was to realize how much this looked like the formula for compound interest. In fact, it *is* the interest formula when you compound 100% interest for 1 unit of time, compounding as fast as possible. The chapter on e describes this interpretation.

- **Definition 1:** Define *e* as 100% compound growth at the smallest increment possible

Let's look at the second definition: an infinite series of terms, getting smaller and smaller. What could this be?

$$e = \frac{1}{0!} + \frac{1}{1!} + \frac{1}{2!} + \frac{1}{3!} + \cdots$$

After noodling this over using the theme of "interest" we see this definition shows the components of compound interest. Now, insights don't come instantly — this insight might strike after brainstorming "What could $1 + 1 + 1/2 + 1/6 + \ldots$" represent when talking about growth?"

Well, the first term ($1 = 1/0!$, remembering that $0!$ is 1) is your principal, the original amount. The next term ($1 = 1/1!$) is the "direct" interest you earned — 100% of 1. The next term ($0.5 = 1/2!$) is the amount of money your interest made ("2nd level interest"). The following term ($.1666 = 1/3!$) is your "3rd-level interest" — how much money your interest's interest earned!

Money earns money, which earns money, which earns money, and so on — the sequence separates out these contributions (the chapter on e describes how Mr. Blue, Mr. Green & Mr. Red grow independently). There's much more to say, but that's the "growth-focused" understanding of that idea.

- **Definition 2:** Define e by the contributions each piece of interest makes

Neato. Now to the third, shortest definition. What does it mean? Instead of thinking "derivative" (which turns your brain into equation-crunching mode), think about what it means. The *feeling* of the equation. Make it your friend.

$$\frac{d}{dx} Blah = Blah$$

It's the calculus way of saying "Your rate of growth is equal to your current amount". Well, growing at your current amount would be a 100% interest rate, right? And by always growing it means you are always calculating interest – it's another way of describing continuously compound interest!

- **Definition 3:** Define e as always growing by 100% of your current value

Nice — e is the number where you're always growing by exactly your current amount (100%), not 1% or 200%.

Time for the last definition — it's a tricky one. Here's my interpretation: Instead of describing how *much* you grew, why not say *how long* it took?

If you're at 1 and growing at 100%, it takes 1 unit of time to get from 1 to 2. But once you're at 2, and growing 100%, it means you're growing at 2 units per unit time! So it only takes $1/2$ unit of time to go from 2 to 3. Going from 3 to 4 only takes $1/3$ unit of time, and so on.

The time needed to grow from 1 to A is the time from 1 to 2, 2 to 3, 3 to 4... and so on, until you get to A. The first definition defines the natural log (ln) as shorthand for this "time to grow" computation.

$ln(a)$ is simply the time to grow from 1 to a. We then say that e is the number that takes exactly 1 unit of time to grow to. Said another way, e is is the amount of growth after waiting exactly 1 unit of time!

- **Definition 4:** Define the time needed to grow continuously from 1 to as $ln(a)$. e is the amount of growth you have after 1 unit of time.

Whablamo! These are four different ways to describe the mysterious e. Once we have the core idea ("e is about 100% continuous growth"), the crazy equations snap into place — it's possible to translate calculus into English. Math is about ideas!

1.4 What's the Moral?

In math class, we often start with the last, most complex idea. It's no wonder we're confused: we're showing students DNA and expecting them to see a cat.

I've learned a few lessons from this approach, and it underlies how I understand and explain math:

- **Search for insights and apply them.** That first intuitive insight can help everything else snap into place. Start with a definition that makes sense and "walk around the circle" to find others.

- **Be resourceful.** Banging your head against an idea is no fun. If it doesn't click, come at it from different angles. There's another book, another article, another person who explains it in a way that makes sense to you.

- **It's ok to be visual.** We think of math as rigid and analytic — but visual interpretations are ok! Do what develops your understanding. Imaginary numbers were puzzling until their geometric interpretation came to light, decades after their initial discovery. Looking at equations all day didn't help mathematicians "get" what they were about.

Math becomes difficult and discouraging when we focus on definitions over understanding. Remember that the modern definition is the most advanced step of thought, not necessarily the starting point. Don't be afraid to approach a concept from a funny angle — figure out the plain-English sentence behind the equation. Happy math.

THE PYTHAGOREAN THEOREM

The Pythagorean theorem ($a^2 + b^2 = c^2$) is a celebrity: if an equation can make it into the Simpsons, I'd say it's well-known.

But most of us think the formula only applies to triangles and geometry. Think again. The Pythagorean Theorem can be used with *any shape* and for *any formula that squares a number*.

Read on to see how this 2500-year-old idea can help us understand computer science, physics, even the value of Web 2.0 social networks.

2.1 Understanding How Area Works

I love seeing old topics in a new light and discovering the depth there. For example, I realize I didn't have a deep grasp of area until writing this chapter. Yes, we can rattle off equations, but do we really *understand* the nature of area? This fact may surprise you:

- The area of any shape can be computed from *any line segment* squared

In a square, our "line segment" is usually a side, and the area is that side squared (side 5, area 25). In a circle, the line segment is often the radius, and the area is πr^2 (radius 5, area 25π). Easy enough.

We can pick any line segment and figure out area from it: every line segment has an "area factor" in this universal equation:

$$Area = Factor \cdot (line\ segment)^2$$

Shape	Image	Line Segment	Area	Area Factor
Square		Side [s]	s^2	1
Square		Perimeter [p]	$\frac{1}{16}p^2$	$\frac{1}{16}$
Square		Diagonal [d]	$\frac{1}{2}d^2$	$\frac{1}{2}$
Circle		Radius [r]	πr^2	$\pi(3.14...)$

For example, look at the diagonal of a square ("d"). A regular side is $d/\sqrt{2}$, so the area becomes $\frac{1}{2}d^2$. Our "area constant" is 1/2 in this case, if we want to use the diagonal as our line segment to be squared.

Now, use the entire perimeter ("p") as the line segment. A side is p/4, so the area is $p^2/16$. The area factor is 1/16 if we want to use p^2.

2.2   Can We Pick Any Line Segment?

You bet. There is always *some* relationship between the traditional line segment (the side of a square), and the one you pick (the perimeter, which happens to be 4 times a side). Since we can convert between the "traditional" and "new" segment, it doesn't matter which one we use – there'll just be a different area factor when we multiply it out.

2.3 Can We Pick Any Shape?

Sort of. A given area formula works for all similar shapes, where "similar" means "zoomed versions of each other". For example:

- All squares are similar (area always s^2)

- All circles are similar, too (area always πr^2)

- All triangles are *not* similar: Some are fat and others skinny – every type of triangle has its own area factor based on the line segment you are using. Change the shape of the triangle and the equation changes.

Yes, every triangle follows the rule $area = \frac{1}{2} base \cdot height$. But the relationship between base and height depends on the type of triangle ($base = 2 \cdot height$, $base = 3 \cdot height$, etc.), so even then the area factor will be different.

Why do we need similar shapes to keep the same area equation? Intuitively, when you zoom (scale) a shape, you're changing the absolute size but not the relative ratios within the shape. A square, no matter how zoomed, has a $perimeter = 4 \cdot side$.

Because the area factor is based on ratios inside the shape, any shapes with the same ratios will follow the same formula. It's a bit like saying everyone's armspan is about equal to their height. No matter if you're a NBA basketball player or child, the equation holds because it's all relative. (This intuitive argument may not satisfy a mathematical mind – in that case, take up your concerns with Euclid).

I hope these high-level concepts make sense:

- Area can be be found from *any line segment squared*, not just the side or radius

- Each line segment has a different "area factor"

- The same area equation works for similar shapes

For the geeks: *why* do all similar shapes have the same area factor? Here's my intuition:

> Scaled versions of the same shape have the same ratios. Why? When we move an object, the apparent size might change (a stop sign close up vs. far away) but it seems the ratios should stay the same. Can an object *know* it's being viewed from afar and modify the ratio of side to area?
>
> Consider two similar shapes. Push away the larger until it's the same apparent size as the smaller. Now they look identical, and therefore have the same ratios (area to perimeter, etc.). Now pull in the larger one. The shape appears bigger, but its ratios haven't changed during the move – they're the same as the smaller shape.

Here's one reader's (Per Vognsen) more formal proof:

You just have to prove that L^2/A is constant within a similarity class. Take two members of the same similarity class, of areas A and A' and lengths L and L'. Let F be the factor of the dilation that maps the first figure onto the other. Then $A = F^2 * A'$ and $L = F * L'$. Squaring the length equation gives $L^2 = F^2 * L'^2$. Dividing the area equation by this, the F^2 factors cancel, yielding $A/L^2 = A'/L'^2$. So the area to squared length ratio is indeed constant.

2.4   Intuitive Look at The Pythagorean Theorem

I think we can all agree the Pythagorean Theorem is true. But most proofs offer a mechanical understanding: re-arrange the shapes, and voila, the equation holds. But is it really clear, intuitively, that it *must* be $a^2 + b^2 = c^2$ and not $2a^2 + b^2 = c^2$? No? Well, let's build some intuition.

There's one killer concept we need: **Any right triangle can be split into two similar right triangles.**

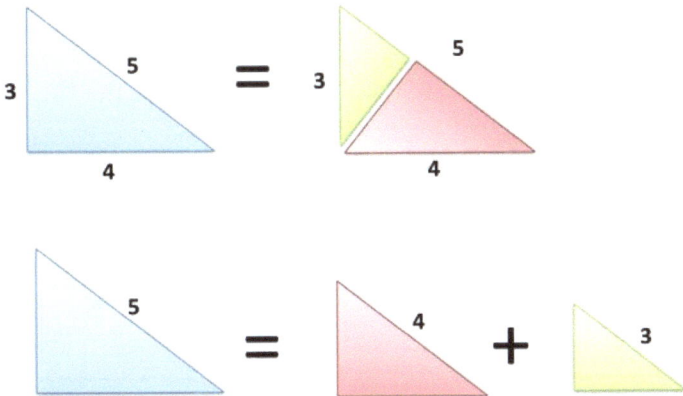

Cool, huh? Drawing a perpendicular line through the point splits a right triangle into two smaller ones. Geometry lovers, try the proof yourself: use angle-angle-angle similarity.

This diagram also makes something very clear:

$$Area(Big) = Area(Medium) + Area(Small)$$

Makes sense, right? The smaller triangles were cut from the big one, so the areas must add up. And the kicker: because the triangles are similar, *they have the same area equation.*

Let's call the long side c (5), the middle side b (4), and the small side a (3). Our area equation for these triangles is:

$$Area = F \cdot hypotenuse^2$$

where F is some area factor (6/25 or .24 in this case; the exact number doesn't matter). Now let's play with the equation:

$$Area(Big) = Area(Medium) + Area(Small)$$
$$F \cdot c^2 = F \cdot b^2 + F \cdot a^2$$

Divide by F on both sides and you get:

$$c^2 = b^2 + a^2$$

Which is our famous theorem! You knew it was true, but now you *know why*:

- A triangle can be split into two smaller, similar ones

- Since the areas must add up, the squared hypotenuses (which determine area) must add up as well

This takes a bit of time to see, but I hope the result is clear. How could the small triangles not add to the larger one?

Actually, it turns out the Pythagorean Theorem depends on the assumptions of Euclidean geometry and doesn't work on spheres or globes, for example. But we'll save that discussion for another time.

2.5   Useful Application: Try Any Shape

We used triangles in our diagram, the simplest 2-D shape. But the line segment can belong to any shape. Take circles, for example:

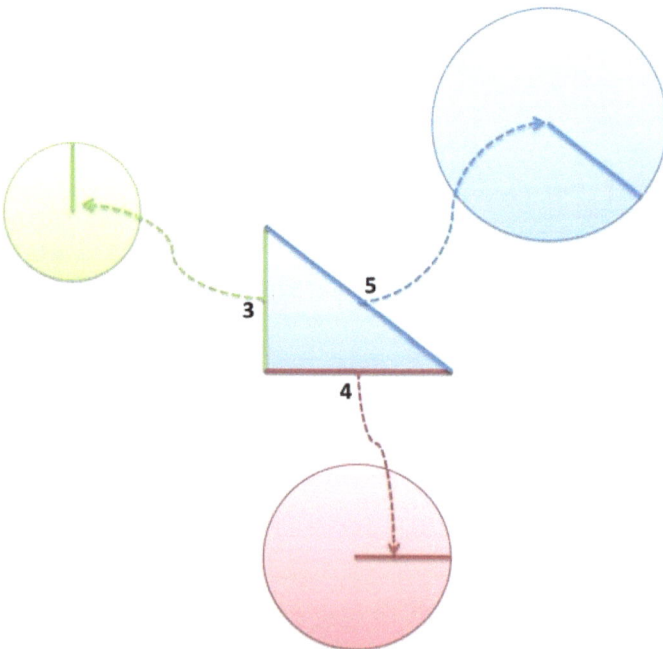

Now what happens when we add them together?

Using Circle Areas

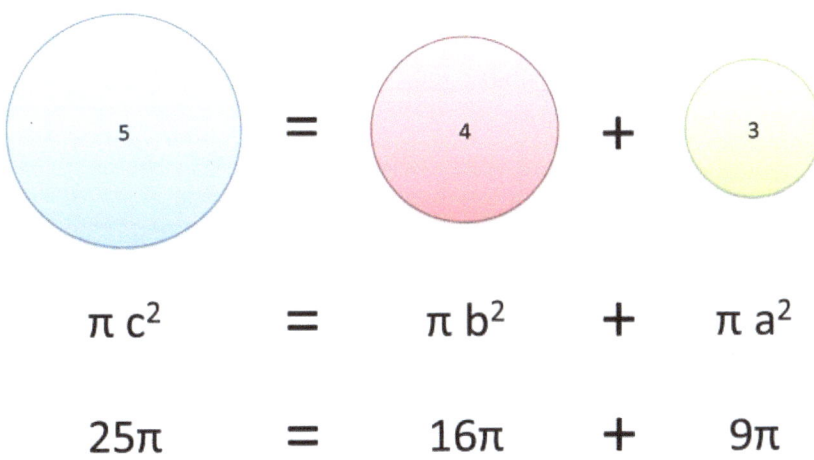

$$\pi\,c^2 \quad = \quad \pi\,b^2 \quad + \quad \pi\,a^2$$

$$25\pi \quad = \quad 16\pi \quad + \quad 9\pi$$

You guessed it: Circle of radius 5 = Circle of radius 4 + Circle of radius 3.

Pretty wild, eh? We can multiply the Pythagorean Theorem by our area factor (π, in this case) and come up with a relationship for any shape.

Remember, the line segment can be *any portion of the shape*. We could have picked the circle's radius, diameter, or circumference – there would be a different area factor, but the 3-4-5 relationship would still hold.

So, whether you're adding up pizzas or Richard Nixon masks, the Pythagorean theorem helps you relate the areas of any similar shapes. Now that's something they didn't teach you in grade school.

2.6 Useful Application: Conservation of Squares

The Pythagorean Theorem applies to *any* equation that has a squared term. The triangle-splitting means you can split any amount (c^2) into two smaller amounts ($a^2 + b^2$) based on the sides of a right triangle. In reality, the "length" of a side can be distance, energy, work, time, or even people in a social network.

Social Networks

Metcalfe's Law (if you believe it) says the value of a network is about n^2 (the number of relationships). In terms of value,

- Network of 50M = Network of 40M + Network of 30M.

Pretty amazing – the 2nd and 3rd networks have 70M people total, but they aren't a coherent whole. The network with 50 million people is as valuable as the others combined.

Computer Science

Some programs with n inputs take n^2 time to run (bubble sort, for example). In terms of processing time:

- 50 inputs = 40 inputs + 30 inputs

Pretty interesting. 70 elements spread among two groups can be sorted as fast as 50 items in one group. (Yeah, there may be constant overhead/start up time, just work with me here).

Given this relationship, it makes sense to partition elements into separate groups and then sort the subgroups. Indeed, that's the approach used in quick-sort, one of the best general-purpose sorting methods. The Pythagorean theorem helps show how sorting 50 combined elements can be as slow as sorting 30 and 40 separate ones.

Surface Area

The surface area of a sphere is $4\pi r^2$. So, in terms of surface area of spheres:

- Area of radius 50 = Area of radius 40 + Area of radius 30

We don't often have spheres lying around, but boat hulls may have the same relationship (they're like deformed spheres, right?). Assuming the boats are similarly shaped, the paint needed to coat one 50 foot yacht could instead paint a 40 and 30-footer. Yowza.

Physics

If you remember your old physics classes, the kinetic energy of an object with mass m and velocity v is $\frac{1}{2}mv^2$. In terms of energy,

- Energy at 500 mph = Energy at 400 mph + Energy at 300 mph

With the energy used to accelerate one bullet to 500 mph, we could accelerate two others to 400 and 300 mph.

2.7 Enjoy Your New Insight

Throughout our school life we think the Pythagorean Theorem is about triangles and geometry. It's not.

When you see a right triangle, realize the sides can represent the lengths of any portion of a shape, and the sides can represent variables in *any equation* that has a square. Maybe it's just me, but I find this pretty surprising.

There's much, much more to this beautiful theorem, such as measuring any distance. Enjoy.

PYTHAGOREAN DISTANCE

We've underestimated the Pythagorean theorem all along. The previous chapter showed it's not about triangles; it can apply to any shape. It's not about a, b and c; it applies to *any formula* with a squared term.

It's not about distance in the sense of walking diagonally across a room. It's about *any distance*, like the "distance" between our movie preferences or colors.

If it can be measured, it can be compared with the Pythagorean Theorem. Let's see why.

3.1   Understanding The Theorem

We agree the theorem works. In any right triangle:

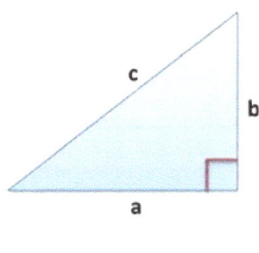

$$a^2 + b^2 = c^2$$

If a=3 and b=4, then c=5. Easy, right? Well, a key observation is that a and b are at right angles (notice the little red box). Movement in one direction has *no impact* on the other.

It's a bit like North/South vs. East/West. Moving North does not change your East/West direction, and vice-versa — the directions are independent (the geek term is orthogonal).

The Pythagorean Theorem lets you find the shortest path distance between orthogonal directions. So it's not really about right "triangles" — it's about comparing "things" moving at right angles.

> **You**: If I walk 3 blocks East and 4 blocks North, how far am I from my starting point?

Me: 5 blocks, as the crow flies. Bring adequate provisions for your journey.
You: Uh, ok.

3.2 So What Is "c"?

Well, we could think of c as just a number, but that keeps us in boring triangle-land. I like to think of c as a *combination* of a and b.

But it's not a simple combination like addition — after all, c doesn't equal a + b. It's more a combination of components — the Pythagorean theorem lets us combine *orthogonal components* in a manner similar to addition. And there's the magic.

In our example, C is 5 blocks of "distance". But it's more than that: it contains a *combination* of 3 blocks East and 4 blocks North. Moving along C means you go East and North at the same time. Neat way to think about it, eh?

3.3 Chaining the Theorem

Let's get crazy and chain the theorem together. Take a look at this:

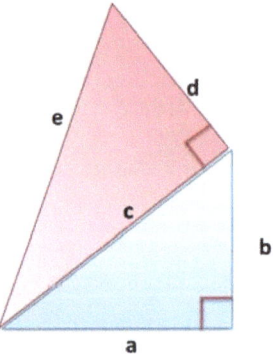

$$a^2 + b^2 = c^2$$
$$c^2 + d^2 = e^2$$

$$\implies \quad a^2 + b^2 + d^2 = e^2$$

Cool, eh? We draw *another* triangle in red, using c as one of the sides. Since c and d are at right angles (orthogonal!), we get the Pythagorean relation: $c^2 + d^2 = e^2$.

And when we replace c^2 with $a^2 + b^2$ we get:

$$a^2 + b^2 + d^2 = e^2$$

And that's something: We've written e in terms of 3 orthogonal components (a, b and d). Starting to see a pattern?

3.4 Put on Your 3D Goggles

Think two triangles are strange? Try pulling one out of the paper. Instead of lining the triangles flat, tilt the red one up:

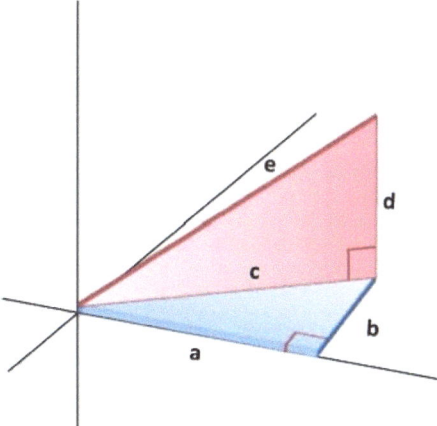

It's the same triangle, just facing a different way. But now we're in 3d! If we call the sides x, y and z instead of a, b and d we get:

$$x^2 + y^2 + z^2 = distance^2$$

Very nice. In math we typically measure the x-coordinate (left/right distance), the y-coordinate (front-back distance), and the z-coordinate (up/down distance). And now we can find the 3-d distance to a point given its coordinates!

3.5   Use Any Number of Dimensions

As you can guess, the Pythagorean Theorem generalizes to *any number of dimensions*. That is, you can chain a bunch of triangles together and tally up the "outside" sections:

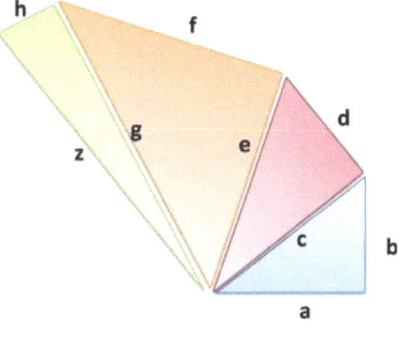

$$a^2 + b^2 + d^2 + f^2 + h^2 = z^2$$

You can imagine that each triangle is in its own dimension. If segments are at right angles, the theorem holds and the math works out.

3.6 How Distance Is Computed

The Pythagorean Theorem is the basis for computing the distance between two points. Consider two triangles:

- Triangle with sides (4,3) [blue]

- Triangle with sides (8,5) [pink]

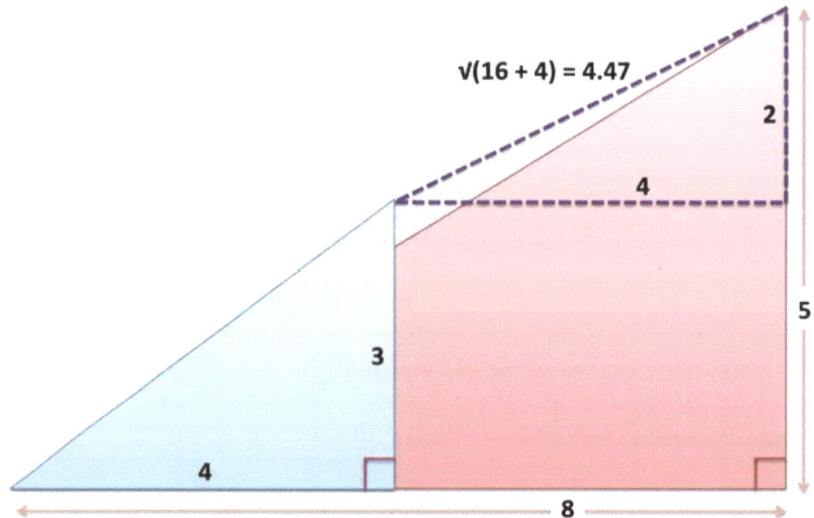

What's the distance from the tip of the blue triangle [at coordinates (4,3)] to the tip of the red triangle [at coordinates (8,5)]? Well, we can create a **virtual triangle** between the endpoints by subtracting corresponding sides. The hypotenuse of the virtual triangle is the distance between points:

- Distance: (8-4,5-3) = (4,2) = sqrt(20) = 4.47

Cool, eh? In 3D, we can find the distance between points (x_1, y_1, z_1) and (x_2, y_2, z_2) using the same approach:

$$distance^2 = (x_2 - x_1)^2 + (y_2 - y_1)^2 + (z_2 - z_1)^2$$

And it doesn't matter if one side is bigger than the other, since the difference is squared and will be positive (another great side-effect of the theorem).

3.7 How to Use Any Distance

The theorem isn't limited to our narrow, spatial definition of distance. It can apply to any orthogonal dimensions: space, time, movie tastes, colors, temperatures. In fact, it can apply to any set of numbers (a,b,c,d,e). Let's take a look.

3.8   Measuring User Preferences

Let's say you do a survey to find movie preferences:

- How did you like Rambo? (1-10)

- How did you like Bambi? (1-10)

- How did you like Seinfeld? (1-10)

How do we compare people's ratings? Find similar preferences? Pythagoras to the rescue!

If we represent ratings as a "point" (Rambo, Bambi, Seinfeld) we can represent our survey responses like this:

- Tough Guy: (10, 1, 3)

- Average Joe: (5, 5, 5)

- Sensitive Guy: (1, 10, 7)

And using the theorem, we can see how "different" people are:

- Tough Guy to Average Joe: (10 – 5, 1 – 5, 3 – 5) = (5, -4, -2) = 6.7

- Tough Guy to Sensitive Guy: (10 – 1, 1 – 10, 3 – 7) = (9, -9, -4) = 13.34

As we suspected, there's a large gap between the Tough and Sensitive Guy, with Average Joe in the middle. The theorem helps us **quantify this distance** and do interesting things like cluster similar results.

This technique can be used to rate Netflix movie preferences and other types of collaborative filtering where you attempt to make predictions based on preferences (i.e. Amazon recommendations). In geek speak, we represented preferences as a vector, and used the theorem to find the distance between them (and group similar items, perhaps).

3.9   Finding Color Distance

Measuring "distance" between colors is another useful application. Colors are represented as red/green/blue (RGB) values from 0(min) to 255 (max). For example

- Black: (0, 0, 0) — no colors

- White: (255, 255, 255) — maximum of each color

- Red: (255, 0, 0) — pure red, no other colors

We can map out all colors in a "color space" like so:

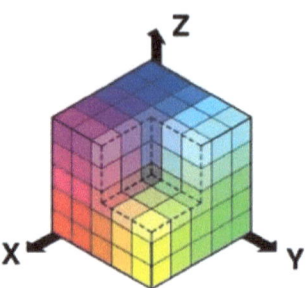

We can compute the distance between colors the usual way: get the distance from our (red, green, blue) value to black (0,0,0). It appears humans can't tell the difference between colors only 4 units apart; heck, even a distance of 30 units looks pretty similar to me:

Red Green Blue	0 0 0	5 5 5	20 20 20	30 30 30	50 50 50	100 100 100
Color						
Distance From Black	0	8.7	34.7	52.0	86.6	173.2

How similar do these look to you? The color distance gives us a **quantifiable** way to measure the distance between colors. You can even unscramble certain blurred images by cleverly applying color distance.

3.10 The Point: You Can Measure Anything

If you can represent a set of characteristics with numbers, you can compare them with the theorem:

- Temperatures during the week: (Mon, Tues, Wed, Thurs, Fri). Compare successive weeks to see how "different" they are (find the difference between 5-dimensional vectors).

- Number of customers coming into a store hour-by-hour, day-by-day, or week-by-week

- Spacetime distance: (latitude, longitude, altitude, date). Useful if you're making a time machine (or a video game that uses one)!

- Differences between people: (Height, Weight, Age)

- Differences between companies: (Revenue, Profit, Market Cap)

You can tweak the distance by weighing traits differently (i.e., multiplying the age difference by a certain factor). But the core idea is so important I'll repeat it again: **If you can quantify it, you can compare it using the Pythagorean Theorem.**

Your x, y and z axes can represent any quantity. And you aren't limited to three dimensions. Sure, mathematicians would love to tell you about the other ways to measure distance (aka metric space), but the Pythagorean Theorem is the most famous and a great starting point.

3.11 So, What Just Happened Here?

There's so much to learn when revisiting concepts we were "taught". Math is beautiful, but the elegance is usually buried under mechanical proofs and a wall of equations. We don't need more proofs; we need interesting, intuitive results.

For example, the Pythagorean Theorem:

- Works for **any shape**, not just triangles (like circles)

- Works for **any equation with squares** (like $\frac{1}{2}mv^2$)

- Generalizes to **any number of dimensions** ($a^2 + b^2 + c^2 + \ldots$)

- Measures **any type of distance** (i.e. between colors or movie preferences)

Not too bad for a 2000-year old formula, right? This is quite a brainful, so I'll finish here for now. Happy math.

RADIANS AND DEGREES

It's an obvious fact that circles should have 360 degrees. Right?

Wrong. Most of us have *no idea* why there's 360 degrees in a circle. We memorize a magic number as the "size of a circle" and set ourselves up for confusion when studying advanced math or physics, with their so called "radians".

"Radians make math easier!" the experts say, without a simple reason why (discussions involving Taylor series are not simple). Today we'll uncover what radians really are, and the intuitive reason they make math easier.

4.1   Where do Degrees Come From?

Before numbers and language we had the stars. Ancient civilizations used astronomy to mark the seasons, predict the future, and appease the gods (when making human sacrifices, *they'd better be on time*).

How is this relevant to angles? Well, bub, riddle me this: isn't it strange that **a circle has 360 degrees and a year has 365 days?**. And isn't it weird that constellations just happen to circle the sky during the course of a year?

Unlike a pirate, I bet you landlubbers can't determine the seasons by the night sky. Here's the Big Dipper (Great Bear) as seen from New York City in 2008:

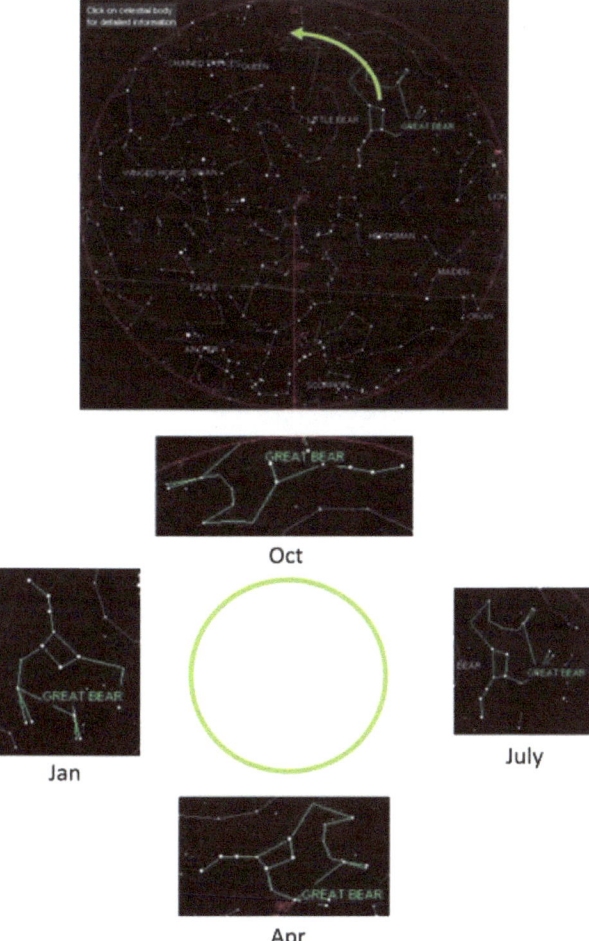

Constellations make a circle every day. If you look at them the same time every day (midnight), they will make a circle throughout the year. Here's a theory about how degrees came to pass:

- Humans noticed that constellations moved in a full circle every year

- Every day, they were off by a tiny bit ("a degree")

- Since a year has about 360 days, a circle had 360 degrees

But, but... why not *365 degrees* in a circle?

Cut 'em some slack: they had *sundials* and didn't know a year should have a convenient 365.242199 degrees like you do.

360 is close enough for government work. It fits nicely into the Babylonian base-60 number system, and divides well (by 2, 3, 4, 6, 10, 12, 15, 30, 45, 90... you get the idea).

4.2 Basing Mathematics on the Sun Seems Perfectly Reasonable

Earth lucked out: ~360 is a great number of days to have in a year. But it does seem arbitrary: on Mars we'd have roughly ~680 degrees in a circle, for the longer Martian year (Martian days are longer too, but you get the idea). And in parts of Europe they've used gradians, where you divide a circle into 400 pieces.

Many explanations stop here saying, "Well, the degree is arbitrary but we need to pick *some* number." Not here: we'll see that **the entire premise of the degree is backwards**.

4.3 Radians Rule, Degrees Drool

A degree is the amount I, an observer, need to tilt my head to see you, the mover. It's a tad self-centered, don't you think?

Suppose you saw a friend go running on a large track:

You: Hey Bill, how far did you go?
Bill: Well, I had a really good pace, I think I went 6 or 7 mile–
You: Shuddup. How far did I turn my head to see you move?
Bill: What?
You: I'll use small words for you. Me in middle of track. You ran around. How... much... did... I turn... my... head?
Bill: Jerk.

Selfish, right? **That's how we do math!** We write equations in terms of "Hey, how far did I turn my head see that planet/pendulum/wheel move?". I bet you've never bothered to think about the pendulum's feelings, hopes and dreams.

Do you think the equations of physics should be made simple for the mover or observer?

4.4 Radians: The Unselfish Choice

Much of physics (and life!) involves leaving your perspective and seeing things from another viewpoint. Instead of wondering how far we tilted our heads, **consider how far the other person moved**.

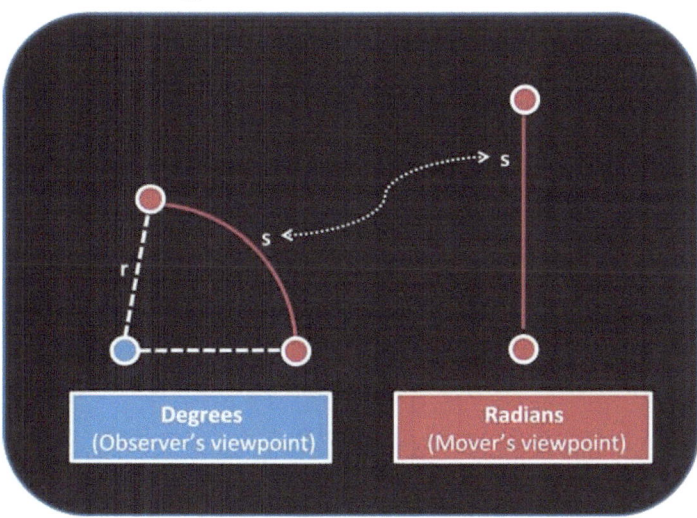

Degrees measure angles by how far we tilted our heads. Radians measure angles by **distance traveled**.

But absolute distance isn't that useful, since going 10 miles is a different number of laps depending on the track. So we divide by radius to get a normalized angle:

$$Radian = \frac{distance\ \ traveled}{radius}$$

You'll often see this as $\theta = s/r$, or angle in radians = arc length divided by radius.

A circle has 360 degrees or 2π radians — going all the way around is $2\pi r/r$. So a radian is about $360/2\pi$ or 57.3 degrees.

Now don't be like me, memorizing this thinking "Great, another unit. 57.3 degrees is so weird." Because it is weird when you're still thinking about you!

Moving 1 radian (unit) is a perfectly normal distance to travel. Put another way, our idea of a "clean, 90 degree angle" means the mover goes a very unclean $\pi/2$ units. Think about it — *"Hey Bill, can you run 90 degrees for me? What's that? Oh, yeah, that'd be $\pi/2$ miles from your point of view."* The strangeness goes both ways.

Radians are the empathetic way to do math — a shift from away from head tilting and towards the mover's perspective.

Strictly speaking, radians are a ratio (length divided by another length) and don't have a dimension. Practically speaking, we're not math robots, and it helps to think of radians as "distance traveled on a unit circle".

4.5 Using Radians

I'm still getting used to thinking in radians. But we encounter the concept of "mover's distance" quite a bit:

- We use "rotations per minute" not "degrees per second" when measuring certain rotational speeds. This is a shift towards the mover's reference point ("How many laps has it gone?") and away from an arbitrary degree measure.

- When a satellite orbits the Earth, we understand its speed in "miles per hour", not "degrees per hour". Now divide by the distance to the satellite and you get the orbital speed in radians per hour.

- Sine, that wonderful function, is defined in terms of *radians* as

$$sin(x) = x - \frac{x^3}{3!} + \frac{x^5}{5!} - \frac{x^7}{7!} \ldots$$

This formula only works when x is in radians! Why? Well, sine is fundamentally related to *distance moved*, not head-tilting. But we'll save that discussion for another day.

4.6   Radian Example 1: Wheels of the Bus

Let's try a real example: you have a bus with wheels of radius 2 meters (it's a monster truck bus). I'll say how fast the wheels are turning and you say how fast the bus is moving. Ready?

"The wheels are turning 2000 degrees per second". You'd think:

- Ok, the wheels are going 2000 degrees per second. That means it's turning 2000/360 or 5 and 5/9ths rotations per second. Circumference = $2\pi r$, so it's moving, um, 2 times 3.14 times 5 and 5/9ths...where's my calculator...

"The wheels are turning 6 radians per second". You'd think:

- Radians are distance along a unit circle — we just scale by the real radius to see how far we've gone. $6 \cdot 2 = 12$ meters per second. Next question.

Wow! No crazy formulas, no π floating around — just *multiply* to convert rotational speed to linear speed. All because radians speak in terms of the mover.

The reverse is easy too. Suppose you're cruising 90 feet per second on the highway (60 miles per hour) on your 24" inch rims (radius 1 foot). How fast are the wheels turning?

Well, 90 feet per second / 1 foot radius = 90 radians per second.

That was easy. I suspect rappers sing about 24" rims for this very reason.

4.7   Radian Example 2: sin(x)

Time for a beefier example. Calculus is about many things, and one concern is what happens when numbers get really big or really small.

Choose a number of degrees (x), and put sin(x) into your calculator:

When you make x small, like .01, $sin(x)$ gets small as well. And the ratio of $sin(x)/x$ seems to be about .017 — what does that mean? Even stranger, what does it mean to multiply or divide by a degree? Can you have square or cubic degrees?

Radians to the rescue! Knowing they refer to distance traveled (they're not just a ratio!), we can interpret the equation this way:

- x is how far you traveled along a circle

- $sin(x)$ is how high on the circle you are

So $sin(x)/x$ is the ratio of how high you are to how far you've gone: the amount of energy that went in an "upward" direction. If you move vertically, that ratio is 100%. If you move horizontally, that ratio is 0%.

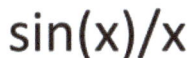

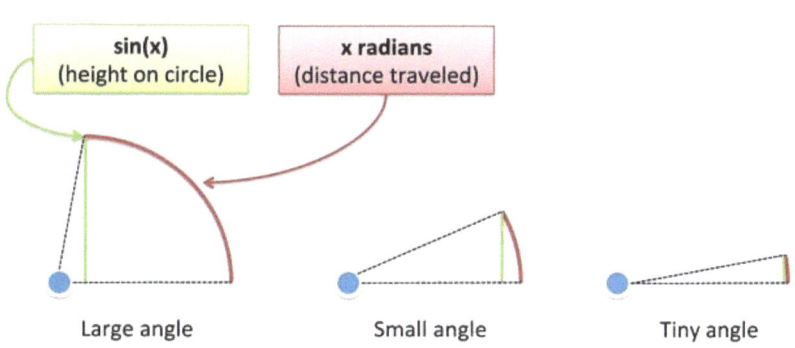

When something moves a tiny amount, such as 0 to 1 degree from our perspective, it's basically going straight up. If you go an even smaller amount, from 0 to .00001 degrees, it's *really* going straight up. The distance traveled (x) is very close to the height ($sin(x)$).

As x shrinks, the ratio gets closer to 100% — more motion is straight up. Radians help us see, intuitively, why $sin(x)/x$ approaches 1 as x gets tiny. We're just nudging along a tiny amount in a vertical direction. By the way, this also explains why $sin(x) \sim x$ for small numbers.

Sure, you can rigorously *prove* this using calculus, but the radian intuition helps you *understand* it.

Remember, these relationships only work when measuring angles with radians. With degrees, you're comparing your height on a circle ($sin(x)$) with how far some observer tilted their head (x degrees), and it gets ugly fast.

4.8   So What's the Point?

Degrees have their place: in our own lives, we're the focal point and want to see how things affect us. How much do I tilt my telescope, spin my snowboard, or turn my steering wheel?

With natural laws, we're an observer describing the motion of others. Radians are about them, not us. It took me many years to realize that:

- Degrees are *arbitrary* because they're based on the sun (365 days ~ 360 degrees), but they are *backwards* because they are from the observer's perspective.

- Because radians are in terms of the mover, equations "click into place". Converting rotational to linear speed is easy, and ideas like $sin(x)/x$ make sense.

Even angles can be seen from more than one viewpoint, and understanding radians makes math and physics equations more intuitive. Happy math.

IMAGINARY NUMBERS

Imaginary numbers always confused me. Like understanding e, most explanations fell into one of two categories:

- It's a mathematical abstraction, and the equations work out. Deal with it.

- It's used in advanced physics, trust us. Just wait until college.

Gee, what a great way to encourage math in kids! Today we'll assault this topic with our favorite tools:

- **Focusing on relationships**, not mechanical formulas.

- **Seeing complex numbers as an upgrade to our number system**, just like zero, decimals and negatives were.

- **Using visual diagrams**, not just text, to understand the idea.

And our secret weapon: learning by analogy. We'll approach imaginary numbers by observing its ancestor, the negatives. Here's your guidebook:

Fun Fact	Negative Numbers (-x)	Complex Numbers (a +bi)
Invented to answer	"What is 3 – 4?"	"What is sqrt(-1)?"
Strange because...	*How can you have less than nothing?*	*How can you take the square root of less than nothing?*
Intuitive meaning	"Opposite"	"Rotation"
Considered absurd until	1700s	Today ☺
Multiplication cycle [& general pattern]	1, −1, 1, −1… X, −X, X, −X…	1, i, −1, −i… X, Y, −X, −Y…
Use in coordinates	Go backwards from origin	Rotate around origin
Measure size with	Absolute value $\sqrt{(-x)^2}$	Pythagorean Theorem $\sqrt{a^2 + b^2}$

It doesn't make sense yet, but hang in there. By the end we'll hunt down *i* and put it in a headlock, instead of the reverse.

5.1 Really Understanding Negative Numbers

Negative numbers aren't easy. Imagine you're a European mathematician in the 1700s. You have 3 and 4, and know you can write 4 – 3 = 1. Simple.

But what about 3-4? What, exactly, does that mean? How can you take 4 cows from 3? *How could you have less than nothing?*

Negatives were considered absurd, something that "darkened the very whole doctrines of the equations" (Francis Maseres, 1759). Yet today, it'd be absurd to think negatives aren't logical or useful. Try asking your teacher whether negatives corrupt the very foundations of math.

What happened? We invented a *theoretical number that had useful properties*. Negatives aren't something we can touch or hold, but they describe certain relationships well (like debt). It was a useful fiction.

Rather than saying "I owe you 30" and reading words to see if I'm up or down, I can write "-30" and know it means I'm in the hole. If I earn money and pay my debts (-30 + 100 = 70), I can record the transaction easily. I have +70 afterwards, which means I'm in the clear.

The positive and negative signs **automatically keep track of the direction** — you don't need a sentence to describe the impact of each transaction. Math became easier, more elegant. It didn't matter if negatives were "tangible" — they had useful properties, and we used them until they became everyday items. Today you'd call someone obscene names if they didn't "get" negatives.

But let's not be smug about the struggle: negative numbers were a huge mental shift. Even Euler, the genius who discovered e and much more, didn't understand negatives as we do today. They were considered "meaningless" results (he later made up for this in style).

It's a testament to our mental potential that today's children are *expected* to understand ideas that once confounded ancient mathematicians.

5.2 Enter Imaginary Numbers

Imaginary numbers have a similar story. We can solve equations like this all day long:

$$x^2 = 9$$

The answers are 3 and -3. But suppose some wiseguy puts in a teensy, tiny minus sign:

$$x^2 = -9$$

Uh oh. This question makes most people cringe the first time they see it. *You want the square root of a number less than zero? That's absurd!*

It seems crazy, just like negatives, zero, and irrationals (non-repeating numbers) must have seemed crazy at first. There's no "real" meaning to this question, right?

Wrong. So-called "imaginary numbers" are as normal as every other number (or just as fake): they're a tool to describe the world. In the same spirit of assuming -1, .3, and 0 "exist", let's assume some number *i* exists where:

$$i^2 = -1$$

That is, you multiply *i* by itself to get -1. What happens now?

Well, first we get a headache. But playing the "Let's pretend *i* exists" game actually makes math easier and more elegant. New relationships emerge that we can describe with ease.

You may not believe in *i*, just like those fuddy old mathematicians didn't believe in -1. New, brain-twisting concepts are **hard** and they don't make sense immediately, even for Euler. But as the negatives showed us, strange concepts can still be useful.

I dislike the term "imaginary number" — it was considered an insult, a slur, designed to hurt *i*'s feelings. The number i is just as normal as other numbers, but the name "imaginary" stuck, so we'll use it.

5.3 Visual Understanding of Negative and Complex Numbers

The equation $x^2 = 9$ really means this:

$$1 \cdot x^2 = 9$$

What transformation x, when applied twice, turns 1 to 9?

The two answers are "x = 3" and "x = -3": That is, you can "scale by" 3 or "scale by 3 and flip" (flipping or taking the opposite is one interpretation of multiplying by a negative).

Now let's think about $x^2 = -1$, which is really

$$1 \cdot x^2 = -1$$

What transformation x, when applied twice, turns 1 into -1? Hrm.

- We can't multiply by a positive twice, because the result stays positive

- We can't multiply by a negative twice, because the result will flip back to positive on the second multiplication

But what about... a **rotation**! It sounds crazy, but if we imagine x being a "rotation of 90 degrees", then applying x twice will be a 180 degree rotation, or a flip from 1 to -1!

Rotate 1 to -1

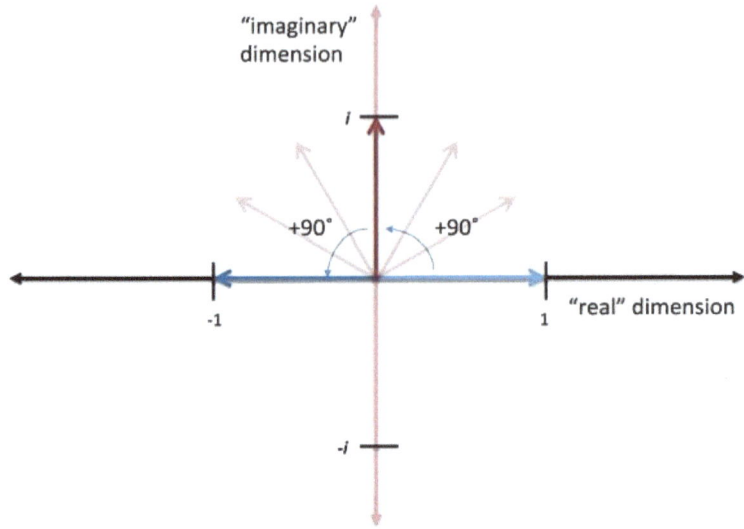

Yowza! And if we think about it more, we could rotate twice in the other direction (clockwise) to turn 1 into -1. This is "negative" rotation or a multiplication by -i:

Positive & Negative Rotation

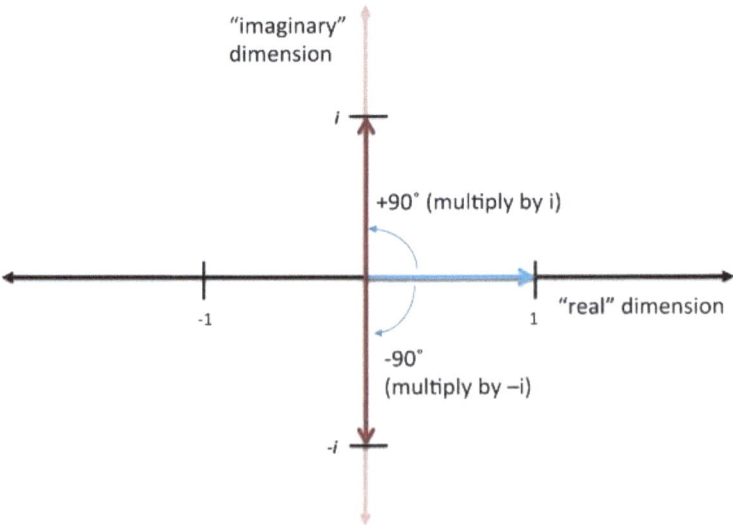

If we multiply by -i twice, we turn 1 into -i, and -i into -1. So there's really *two* square roots of -1: i and $-i$.

This is pretty cool. We have some sort of answer, but what does it mean?

- i is a "new imaginary dimension" to measure a number

- i (or -i) is what numbers "become" when rotated

- Multiplying i is a rotation by 90 degrees counter-clockwise

- Multiplying by -i is a rotation of 90 degrees clockwise

- Two rotations in either direction is -1: it brings us back into the "regular" dimensions of positive and negative numbers.

Numbers are 2-dimensional. Yes, it's mind bending, just like decimals or long division would be mind-bending to an ancient Roman (*What do you mean there's a number between 1 and 2?*).

We asked "How do we turn 1 into -1 in two steps?" and found an answer: rotate it 90 degrees. It's a strange, new way to think about math. But it's useful. (By the way, this geometric interpretation of complex numbers didn't arrive until decades after i was discovered).

Also, keep in mind that having counter-clockwise be positive is a human convention — it easily could have been the other way.

5.4 Finding Patterns

Let's dive into the details a bit. When multiplying negative numbers (like -1), you get a pattern:

$$1, -1, 1, -1, 1, -1, 1, -1\ldots$$

Since -1 doesn't change the **size** of a number, just the sign, you flip back and forth. For some number "x", you'd get:

$$x, -x, x, -x, x, -x\ldots$$

This idea is useful. The number "x" can represent a good or bad hair week. Suppose weeks alternate between good and bad; this is a good week; what will it be like in 47 weeks?

$$x \cdot -1^{47} = x \cdot -1 = -x$$

So -x means a bad hair week. Notice how negative numbers "keep track of the sign" — we can throw -1^{47} into a calculator without having to count ("*Week 1 is good, week 2 is bad... week 3 is good...*"). Things that **flip back and forth can be modeled well with negative numbers.**

Ok. Now what happens if we keep multiplying by i?

$$1, i, i^2, i^3, i^4, i^5\ldots$$

Very funny. Let's reduce this a bit:

- $1 = 1$ (No questions here)

- $i = i$ (Can't do much)

- $i^2 = -1$ (That's what i is all about)

- $i^3 = (i \cdot i) \cdot i = -1 \cdot i = -i$ (Ah, 3 rotations counter-clockwise = 1 rotation clockwise. Neat.)

- $i^4 = (i \cdot i) \cdot (i \cdot i) = -1 \cdot -1 = 1$ (4 rotations bring us "full circle")

- $i^5 = i^4 \cdot i = 1 \cdot i = i$ (Here we go again...)

Represented visually:

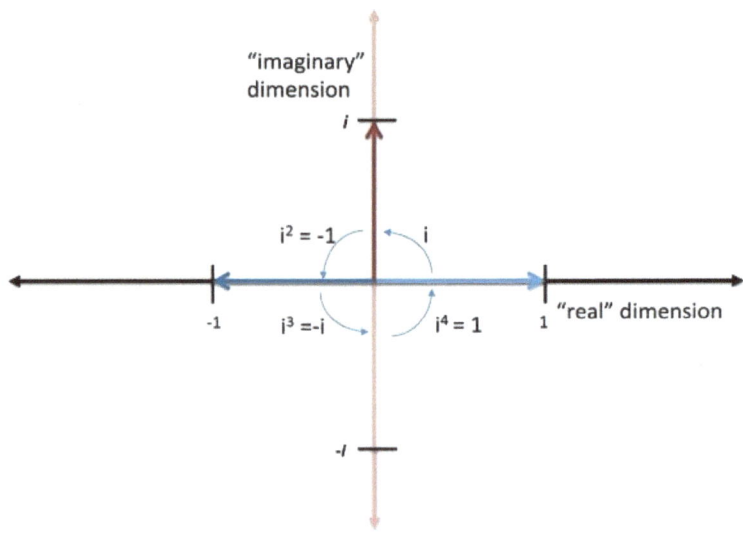

We cycle every 4th rotation. This makes sense, right? Any kid can tell you that 4 left turns is the same as no turns at all. Now rather than focusing on imaginary numbers (i, i^2), look at the general pattern:

$$X, Y, -X, -Y, X, Y, -X, -Y \ldots$$

Like negative numbers modeling flipping, imaginary numbers **can model anything that rotates** between two dimensions "X" and "Y". Or anything with a cyclic, circular relationship — have anything in mind?

'Cos it'd be a sin if you didn't. There'll de Moivre be more in the last chapter. *[Editor's note: Kalid is in electroshock therapy to treat his pun addiction.]*

5.5 Understanding Complex Numbers

There's another detail to cover: can a number be both "real" and "imaginary"?

You bet. Who says we have to rotate the entire 90 degrees? If we keep 1 foot in the "real" dimension and another in the imaginary one, it looks like this:

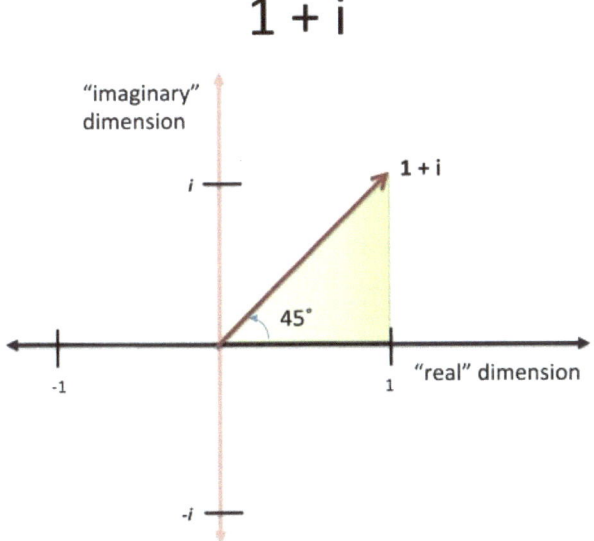

We're at a 45 degree angle, with equal parts in the real and imaginary (1 + i). It's like a hotdog with both mustard and ketchup — who says you need to choose?

In fact, we can pick any combination of real and imaginary numbers and make a triangle. The angle becomes the "angle of rotation". A **complex number** is the fancy name for numbers with both real and imaginary parts. They're written a + bi, where

- a is the real part

- b is the imaginary part

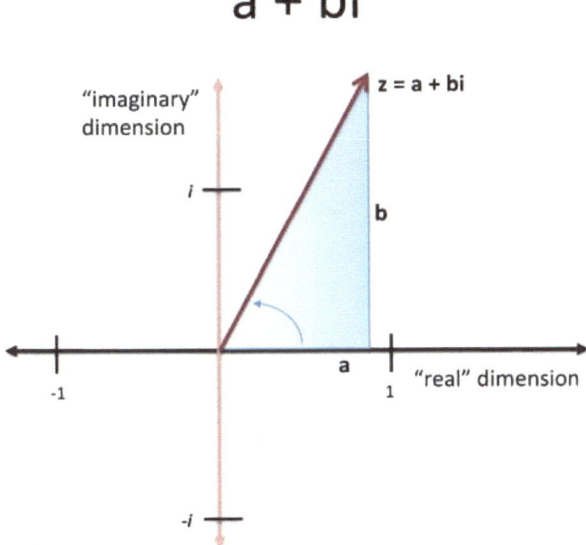

Not too bad. But there's one last question: how "big" is a complex number? We can't measure the real part or imaginary parts in isolation, because that would miss the big picture.

Let's step back. The size of a negative number is not whether you can count it — it's the distance from zero. In the case of negatives this is:

$$Size \; of \; (-x) = \sqrt{(-x)^2} = |x|$$

Which is another way to find the absolute value. But for complex numbers, how do we measure two components at 90 degree angles?

It's a bird... it's a plane... it's Pythagoras!

Geez, his theorem shows up everywhere, even in numbers invented 2000 years after his time. Yes, we are making a triangle of sorts, and the hypotenuse is the distance from zero:

$$Size \; of \; a + bi = \sqrt{a^2 + b^2}$$

Neat. While measuring the size isn't as easy as "dropping the negative sign", complex numbers do have their uses. Let's take a look.

5.6 A Real Example: Rotations

We're not going to wait until college physics to use imaginary numbers. Let's try them out *today*. There's much more to say about complex multiplication, but keep this in mind:

- Multiplying by a complex number rotates by its angle

Let's take a look. Suppose I'm on a boat, with a heading of 3 units East for every 4 units North. I want to change my heading 45 degrees counterclockwise. What's the new heading?

Find the heading

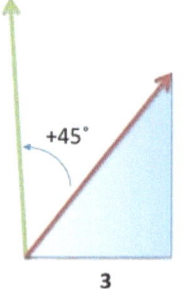

Some hotshot will say *"That's simple! Just take the sine, cosine, gobbledegook by the tangent...fluxsom the foobar...and..."*. **Crack.** Sorry, did I break your calculator? Care to answer that question again?

Let's try a simpler approach: we're on a heading of 3 + 4i (whatever that angle is; we don't really care), and want to rotate by 45 degrees. Well, 45 degrees is 1 + i, so we can multiply by that amount!

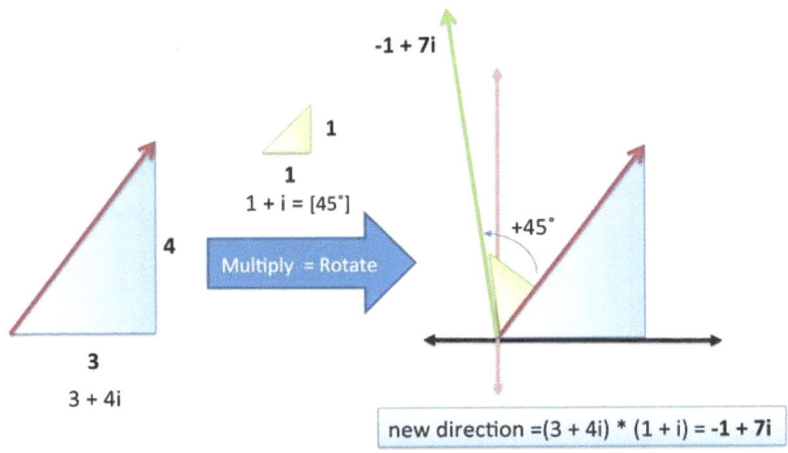

Applying Complex Numbers

Here's the idea:

- Original heading: 3 units East, 4 units North = 3 + 4i

- Rotate counter-clockwise by 45 degrees = multiply by 1 + i

If we multiply them together we get:

$$(3+4i) \cdot (1+i) = 3+4i+3i+4i^2 = 3-4+7i = -1+7i$$

So our new orientation is 1 unit West (-1 East), and 7 units North, which you could draw out and follow.

But yowza! We found that out in 10 seconds, without touching sine or cosine. There were no vectors, matrices, or keeping track of what quadrant we are in. It was **just arithmetic** with a touch of algebra to cross-multiply. Imaginary numbers have the rotation rules baked in: **it just works.**

Even better, the result is useful. We have a heading (-1, 7) instead of an angle (atan(7/-1) = 98.13, keeping in mind we're in quadrant 2). How, exactly, were you planning on drawing and following that angle? With the protractor you keep around?

No, you'd convert it into cosine and sine (-.14 and .99), find a reasonable ratio between them (about 1 to 7), and sketch out the triangle. Complex numbers beat you to it, instantly, accurately, and without a calculator.

If you're like me, you'll find this use **mind-blowing**. And if you don't, well, I'm afraid math doesn't toot your horn. Sorry.

Trigonometry is great, but complex numbers can make ugly calculations simple (like calculating cosine(a+b)). This is just a preview; the next chapter will give you the full meal.

5.7 Complex Numbers Aren't

That was a whirlwind tour of my basic insights. Take a look at the first chart — it should make sense now.

There's so much more to these beautiful, zany numbers, but my brain is tired. My goals were simple:

- Convince you that complex numbers were considered "crazy" but can be useful (just like negative numbers were)

- Show how complex numbers can make certain problems easier, like rotations

If I seem hot and bothered about this topic, there's a reason. Imaginary numbers have been a bee in my bonnet for years — the lack of an intuitive insight frustrated me.

Now that I've finally had insights, I'm bursting to share them. We often suffocate our questions and "chug through", which leaves us with a very fragile understanding. These revelations are my little candle in the darkness; you'll shine a spotlight of your own.

There's much more to complex numbers: check out complex arithmetic in the next chapter. Happy math.

5.8 Epilogue: But They're Still Strange!

I know, they're still strange to me too. I try to put myself in the mind of the first person to discover zero.

Zero is such a weird idea, having "something" represent "nothing", and it eluded the Romans. Complex numbers are similar — it's a new way of thinking. But both zero and complex numbers make math much easier. If we never adopted strange, new number systems, we'd still be counting on our fingers.

I repeat this analogy because it's **so easy** to start thinking that complex numbers aren't "normal". Let's keep our mind open: in the future they'll chuckle that complex numbers were once distrusted, even until the 2000's.

6

COMPLEX ARITHMETIC

Imaginary numbers have an intuitive explanation: they "rotate" numbers, just like negatives make a "mirror image" of a number. This insight makes arithmetic with complex numbers easier to understand, and is a great way to double-check your results. Here's our cheatsheet:

Complex Operation	Intuitive Meaning
Magnitude: $\|z\|$	Distance from zero: $\|z\| = \sqrt{a^2 + b^2}$
Addition & Subtraction	Sliding numbers
Multiplication	Scale by magnitude, add angles
Division	Shrink by magnitude, subtract angles
Complex Conjugate: z*	"Imaginary Reflection": Same size, opposite angle If z = 3 + 4i , then z*= 3 − 4i
Conjugate Properties	(x + y)* = x* + y* [add then reflect = reflect then add] (xy)* = x*y* [multiply then reflect = reflect then multiply]

This chapter will walk through the intuitive meanings.

6.1 Complex Variables

In regular algebra, we often say "x = 3" and all is dandy — there's some number "x", whose value is 3. With complex numbers, there's a gotcha: there's two dimensions to talk about. When writing

$$z = 3 + 4i$$

a + bi

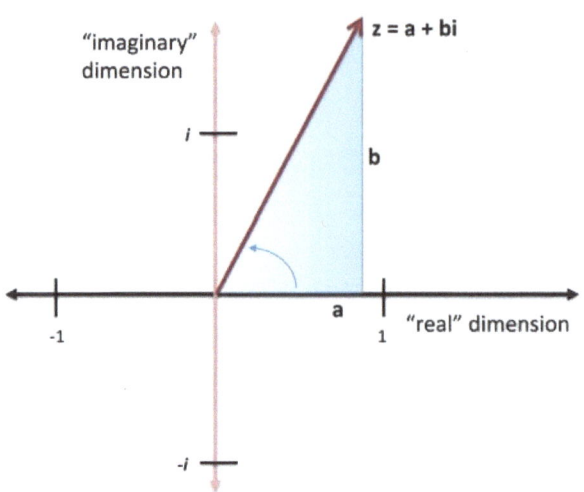

we're saying there's a number "z" with two parts: 3 (the real part) and 4i (imaginary part). It is a bit strange how "one" number can have two parts, but we've been doing this for a while. We often write:

$$y = 3\frac{4}{10} = 3 + .4$$

and it doesn't bother us that a single number "y" has both an integer part (3) and a fractional part (.4 or 4/10). Y is a combination of the two. Complex numbers are similar: they have their real and imaginary parts "contained" in a single variable (shorthand is often Re and Im).

Unfortunately, we don't have nice notation like (3.4) to "merge" the parts into a single number. I had an idea to write the imaginary part vertically, in fading ink, but it wasn't very popular. So we'll stick to the "a + bi" format.

6.2 Measuring Size

Because complex numbers use two independent axes, we find size (magnitude) using the Pythagorean Theorem:

Magnitude of a + bi

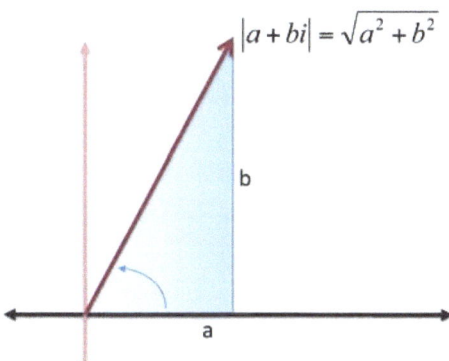

$$|a + bi| = \sqrt{a^2 + b^2}$$

So, a number $z = 3 + 4i$ would have a magnitude of 5. The shorthand for "magnitude of z" is this: $|z|$

See how it looks like the absolute value sign? Well, in a way, it is. Magnitude measures a complex number's "distance from zero", just like absolute value measures a negative number's "distance from zero".

6.3   Complex Addition and Subtraction

We've seen that regular addition can be thought of as "sliding" by a number. Addition with complex numbers is similar, but we can slide in two dimensions (real or imaginary). For example:

Complex Addition

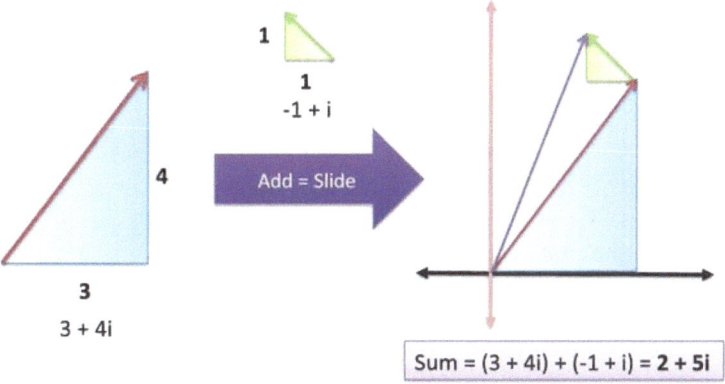

Sum = (3 + 4i) + (-1 + i) = **2 + 5i**

Adding $(3 + 4i)$ to $(-1 + i)$ gives $2 + 5i$.

Again, this is a visual interpretation of how "independent components" are combined: we track the real and imaginary parts separately.

Subtraction is the reverse of addition — it's sliding in the *opposite* direction. Subtracting $(1 + i)$ is the same as adding $-1 \cdot (1 + i)$, or adding $(-1 - i)$.

6.4   Complex Multiplication

Here's where the math gets interesting. When we multiply two complex numbers (x and y) to get z:

- **Add the angles:** $angle(z) = angle(x) + angle(y)$

- **Multiply the magnitudes:** $|z| = |x| \cdot |y|$

That is, the angle of z is the sum of the angles of x and y, and the magnitude of z is the product of the magnitudes. Believe it or not, the magic of complex numbers makes the math work out!

Multiplying by the magnitude (size) makes sense — we're used to that happening in regular multiplication (3 × 4 means you multiply 3 by 4's size). The reason the angle addition works is more detailed, and we'll save it for another time. (Curious? Find the sine and cosine addition formulas and compare them to how $(a + bi) \cdot (c + di)$ get multiplied out).

Time for an example: let's multiply z = 3 + 4i by itself. Before doing all the math, we know a few things:

- The resulting magnitude will be 25. z has a magnitude of 5, so $|z| \cdot |z| = 25$.

- The resulting angle will be above 90. $3 + 4i$ is above 45 degrees (since $3 + 3i$ would be 45 degrees), so twice that angle will be more than 90.

With our predictions on paper, we can do the math:

$$(3 + 4i) \cdot (3 + 4i) = 9 + 16i^2 + 24i = -7 + 24i$$

Time to check our results:

- Magnitude: $\sqrt{(-7 \times -7) + (24 \times 24)} = \sqrt{625} = 25$, which matches our guess.

- Angle: Since -7 is negative and 24i is positive, we know we are going "backwards and up", which means we've crossed 90 degrees ("straight up"). Getting geeky, we compute atan(24/-7) = 106.2 degrees (keeping in mind we're in quadrant 2). This guess checks out too.

Nice. While we can always do the math out, the intuition about rotations and scaling helps us check the result. If the resulting angle was less than 90 ("forward and up", for example), or the resulting magnitude not 25, we'd know there was a mistake in our math.

6.5   Complex Division

Division is the opposite of multiplication, just like subtraction is the opposite of addition. When dividing complex numbers (x divided by y), we:

- **Subtract angles** $angle(z) = angle(x) - angle(y)$

- **Divide by magnitude** $|z| = |x|/|y|$

Sounds good. Now let's try to do it:

$$\frac{3+4i}{1+i}$$

Hrm. Where to start? How do we actually do the division? Dividing regular algebraic numbers gives me the creeps, let alone weirdness of i (*Mister mister! Didya know that $1/i = -i$? Just multiply both sides by i and see for yourself!* Eek.). Luckily there's a shortcut.

6.6   Introducing Complex Conjugates

Our first goal of division is to subtract angles. How do we do this? Multiply by the opposite angle! This will "add" a negative angle, doing an angle subtraction.

Instead of $z = a + bi$, think about a number $z^* = a - bi$, called the "complex conjugate". It has the same real part, but is the "mirror image" in the imaginary dimension. The conjugate or "imaginary reflection" has the same magnitude, but the opposite angle!

So, multiplying by $a - bi$ is the same as subtracting an angle. Neato.

Complex conjugates are indicated by a star (z^*) or bar ($\bar{z}$) above the number – mathematicians love to argue about these notational conventions. Either way, the conjugate is the complex number with the imaginary part flipped:

$$z = a + bi$$

has the complex conjugate

$$z^* = \bar{z} = a - bi$$

Complex Conjugates

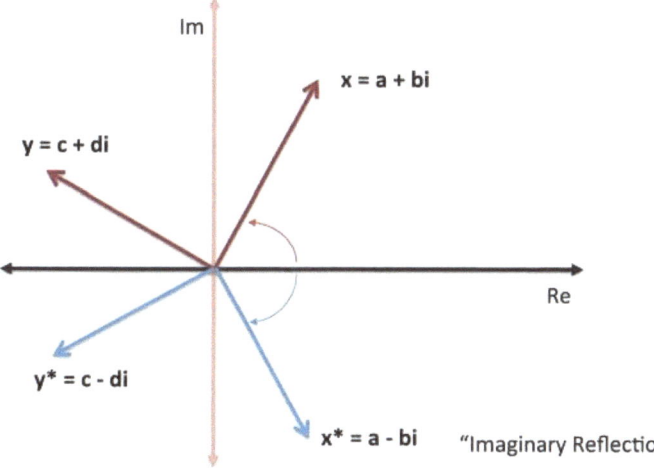

Note that b doesn't have to be "negative". If $z = 3 - 4i$, then $z^* = 3 + 4i$.

6.7 Multiplying By the Conjugate

What happens if you multiply by the conjugate? What is z times z*? Without thinking, think about this:

$$z \cdot z^* = 1 \cdot z \cdot z^*$$

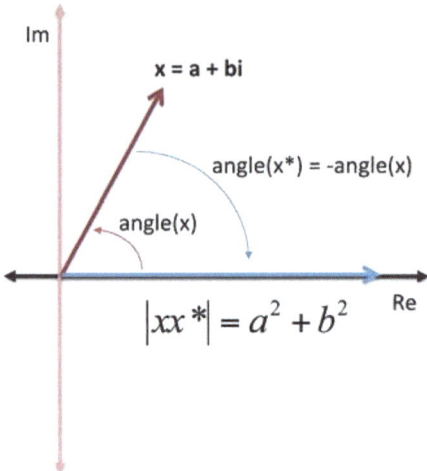

So we take 1 (a real number), add *angle(z)*, and add angle (z^*). But this last angle is negative — it's a subtraction! So our final result should be a real number, since we've canceled the angles. The number should be $|z|^2$ since we scaled by the size twice.

Now let's do an example: $(3 + 4i) \cdot (3 - 4i) = 9 - 16i^2 = 25$

We got a real number, like we expected! The math fans can try the algebra also:

$$(a + bi) \cdot (a - bi) = a^2 + abi - abi - b^2 i^2 = a^2 + b^2$$

Tada! The result has no imaginary parts, and is the magnitude squared. Understanding complex conjugates as a "negative rotation" lets us predict these results in a different way.

6.8 Scaling Your Numbers

When multiplying by a conjugate z^*, we scale by the magnitude $|z^*|$. To reverse this effect we can divide by $|z|$, and to actually *shrink* by $|z|$ we have to divide again. All in all, we have to divide by $|z| \cdot |z|$ to the original number after multiplying by the conjugate.

6.9   Show Me The Division!

I've been sidestepping the division, and here's the magic. If we want to do

$$\frac{3+4i}{1+i}$$

We can approach it intuitively:

- Rotate by opposite angle: multiply by (1 - i) instead of (1 + i)

- Divide by magnitude squared: divide by $|\sqrt{2}|^2 = 2$

The answer, using this approach, is:

$$\frac{3+4i}{1+i} = (3+4i)\cdot(1-i)\cdot\frac{1}{2} = (3-4i^2+4i-3i)\cdot\frac{1}{2} = \frac{7}{2}+\frac{1}{2}i$$

The more traditional "plug and chug" method is to multiply top and bottom by the complex conjugate:

$$\frac{3+4i}{1+i} = \frac{3+4i}{1+i}\cdot\frac{1-i}{1-i} = \frac{3-4i^2+4i-3i}{1-i^2} = \frac{7+i}{2}$$

We're traditionally taught to "just multiply both sides by the complex conjugate" without questioning what complex division really means. But not today.

   We know what's happening: division is subtracting an angle and shrinking the magnitude. By multiplying top and bottom by the conjugate, we subtract by the angle of (1-i), which happens to make the denominator a real number (it's no coincidence, since it's the exact opposite angle). We scaled both the top and bottom by the same amount, so the effects cancel. The result is to turn division into a multiplication in the numerator.

   Both approaches work (you're usually taught the second), but it's nice to have one to double-check the other.

6.10   More Math Tricks

Now that we understand the conjugate, there's a few properties to consider:

$$(x+y)^* = x^* + y^*$$
$$(x\cdot y)^* = x^* \cdot y^*$$

The first should make sense. Adding two numbers and "reflecting" (conjugating) the result, is the same as adding the reflections. Another way to think about it: sliding two numbers then taking the opposite, is the same as sliding both times in the opposite direction.

   The second property is trickier. Sure, the algebra may work, but what's the intuitive explanation? The result $(x\cdot y)^*$ means:

- Multiply the magnitudes: $|x|\cdot|y|$

- Add the angles and take the conjugate (opposite): $angle(x) + angle(y)$ becomes $-angle(x) + -angle(y)$

And x^* times y^* means:

- Multiply the magnitudes: $|x| \cdot |y|$ (this is the same as above)

- Add the conjugate angles: $angle(x^*) + angle(y^*) = -angle(x) + -angle(y)$

Aha! We get the same angle and magnitude in each case, and we didn't have to jump into the traditional algebra explanation. Algebra is fine, but it isn't always the most satisfying explanation.

6.11 A Quick Example

The conjugate is a way to "undo" a rotation. Think about it this way:

- I deposited $3, $10, $15.75 and $23.50 into my account. What transaction will cancel these out? To find the opposite: add them up, and multiply by -1.

- I rotated a line by doing several multiplications: $(3 + 4i)$, $(1 + i)$, and $(2 + 10i)$. What rotation will cancel these out? To find the opposite: multiply the complex numbers together, and take the conjugate of the result.

See the conjugate z^* as a way to "cancel" the rotation effects of z, just like a negative number "cancels" the effects of addition. One caveat: with conjugates, you need to divide by $|z| \cdot |z|$ to remove the scaling effects as well.

6.12 Closing Thoughts

The math here isn't new, but I never realized *why* complex conjugates worked as they did. Why $a - bi$ and not $-a + bi$? Well, complex conjugates are not a random choice, but a mirror image from the imaginary perspective, with the exact opposite angle.

Seeing imaginary numbers as rotations gives us a new mindset to approach problems; the "plug and chug" formulas can make intuitive sense, even for a strange topic like complex numbers. Happy math.

EXPONENTIAL FUNCTIONS & *e*

e has always bothered me — not the letter, but the mathematical constant. What does it really mean?

Math books and even my beloved Wikipedia describe *e* using obtuse jargon:

> *The mathematical constant e is the base of the natural logarithm.*

And when you look up natural logarithm you get:

> *The natural logarithm, formerly known as the hyperbolic logarithm, is the logarithm to the base e, where e is an irrational constant approximately equal to 2.718281828459.*

Nice circular reference there. It's like a dictionary that defines labyrinthine with Byzantine: it's correct but not helpful. What's wrong with everyday words like "complicated"?

I'm not picking on Wikipedia — many math explanations are dry and formal in their quest for "rigor". But this doesn't help beginners trying to get a handle on a subject (and we were all a beginner at one point).

No more! Today I'm sharing my intuitive, high-level insights about what *e* is and why it rocks. Save your "rigorous" math book for another time.

7.1   *e* is Not Just a Number

Describing *e* as "a constant approximately 2.71828..." is like calling pi (π) "an irrational number, approximately equal to 3.1415...". Sure, it's true, but you completely missed the point.

Pi is the ratio between circumference and diameter shared by all circles. It is a fundamental ratio inherent in all circles and therefore impacts any calculation of circumference, area, volume, and surface area for circles, spheres, cylinders, and so on. Pi is important and shows all circles are related, not to mention the trigonometric functions derived from circles (sin, cos, tan).

e **is the base amount of growth shared by all continually growing processes.** *e* lets you take a simple growth rate (where all change happens at the end of the year) and find the impact of compound, continuous growth, where every nanosecond (or faster) you are growing just a little bit.

e shows up whenever systems grow exponentially and continuously: population, radioactive decay, interest calculations, and more. Even jagged systems that don't grow smoothly can be *approximated* by e.

Just like every number can be considered a "scaled" version of 1 (the base unit), every circle can be considered a "scaled" version of the unit circle (radius 1), and every rate of growth can be considered a "scaled" version of e (the "unit" rate of growth).

So e is not an obscure, seemingly random number. e represents the idea that all continually growing systems are scaled versions of a common rate.

7.2 Understanding Exponential Growth

Let start by looking at a basic system that doubles after an amount of time. For example,

- Bacteria can split and "doubles" every 24 hours

- We get twice as many noodles when we fold them in half.

- Your money doubles every year if you get 100% return (lucky!)

And it looks like this:

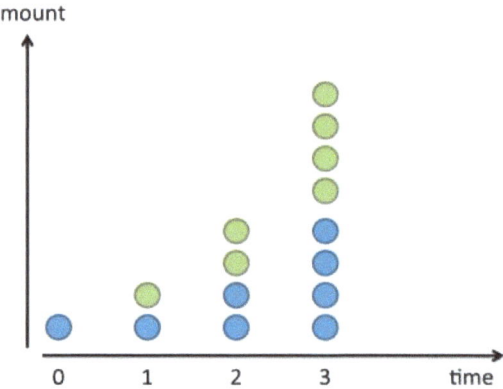

Splitting in two or doubling is a very common progression. Sure, we can triple or quadruple, but doubling is convenient, so hang with me here.

Mathematically, if we have x splits then we get 2^x times more "stuff" than when we started. With 1 split we have 2^1 or 2 times more. With 4 splits we have $2^4 = 16$ times more. As a general formula:

$$growth = 2^x$$

Said another way, doubling is 100% growth. We can rewrite our formula like this:

$$growth = (1 + 100\%)^x$$

It's the same equation, but we separate "2" into what it really is: the original value (1) plus 100%. Clever, eh?

Of course, we can substitute 100% for any number (50%, 25%, 200%) and get the growth formula for that new rate. So the general formula for x periods of return is:

$$growth = (1 + return)^x$$

This just means we multiply by our rate of return (1 + return) x times.

7.3   A Closer Look

Our formula assumes growth happens in discrete steps. Our bacteria are waiting, waiting, and then boom, they double at the very last minute. Our interest earnings magically appear at the 1 year mark. Based on the formula above, growth is punctuated and happens instantly. The green dots suddenly appear.

The world isn't always like this. If we zoom in, we see that our bacterial friends split over time:

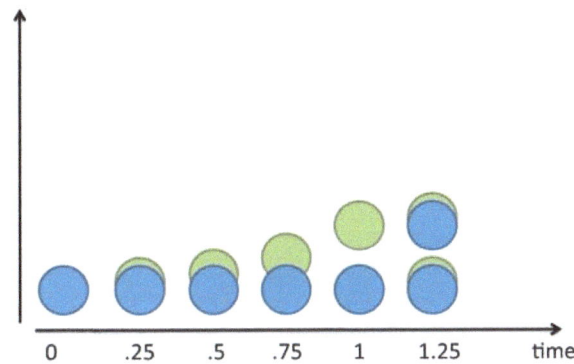

Mr. Green doesn't just show up: he slowly grows out of Mr. Blue. After 1 unit of time (24 hours in our case), Mr. Green is complete. He then becomes a mature blue cell and can create new green cells of his own.

Does this information change our equation?

Nope. In the bacteria case, the half-formed green cells still can't do anything until they are fully grown and separated from their blue parents. The equation still holds.

7.4   Money Changes Everything

But money is different. As soon as we earn a penny of interest, that penny can start earning micro-pennies of its own. We don't need to wait until we earn a complete dollar in interest — fresh money doesn't need to mature.

Based on our *old formula*, interest growth looks like this:

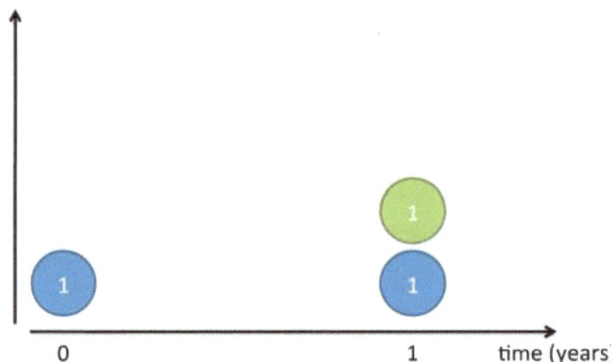

But again, this isn't quite right: all the interest appears on the last day. Let's zoom in and split the year into two chunks. We earn 100% interest every year, or 50% every 6 months. So, we earn 50 cents the first 6 months and another 50 cents in the last half of the year:

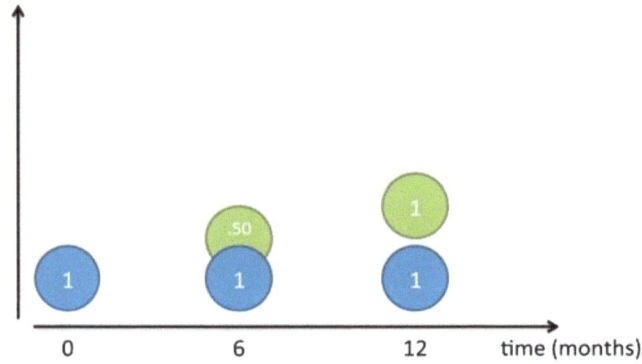

But this still isn't right! Sure, our original dollar (Mr. Blue) earns a dollar over the course of a year. But after 6 months we had a 50-cent piece, ready to go, that we neglected! That 50 cents could have earned money on its own:

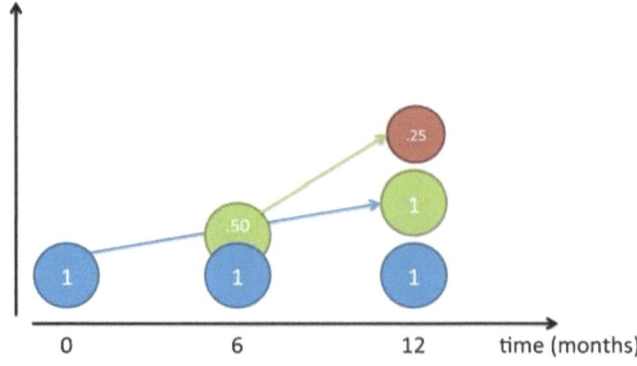

Because our rate is 50% per half year, that 50 cents would have earned 25 cents (50% times 50 cents). At the end of 1 year we'd have

- Our original dollar (Mr. Blue)

- The dollar Mr. Blue made (Mr. Green)

- The 25 cents Mr. Green made (Mr. Red)

Giving us a total of $2.25. We gained $1.25 from our initial dollar, even better than doubling!

Let's turn our return into a formula. The growth of two half-periods of 50% is:

$$growth = (1 + 100\%/2)^2 = 2.25$$

7.5 Diving Into Compound Growth

It's time to step it up a notch. Instead of splitting growth into two periods of 50% increase, let's split it into 3 segments of 33% growth. Who says we have to wait for 6 months before we start getting interest? Let's get more granular in our counting.

Charting our growth for 3 compounded periods gives a funny picture:

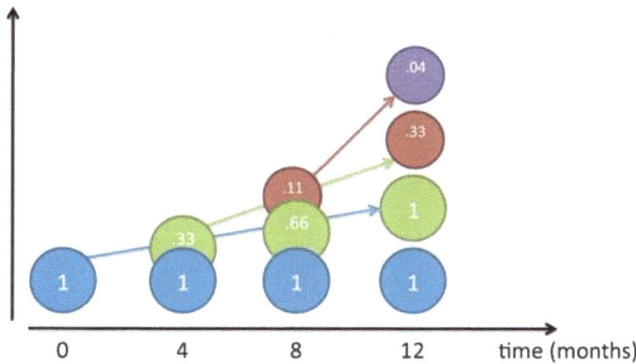

Think of each color as "shoveling" money upwards towards the other colors (its children), at 33% per period:

- **Month 0:** We start with Mr. Blue at $1.

- **Month 4:** Mr. Blue has earned 1/3 dollar on himself, and creates Mr. Green, shoveling along 33 cents.

- **Month 8:** Mr. Blue earns another 33 cents and gives it to Mr. Green, bringing Mr. Green up to 66 cents. Mr. Green has actually earned 33% on his previous value, creating 11 cents (33% × 33 cents). This 11 cents becomes Mr. Red.

- **Month 12:** Things get a bit crazy. Mr. Blue earns another 33 cents and shovels it to Mr. Green, bringing Mr. Green to a full dollar. Mr. Green earns 33% return on his Month 8 value (66 cents), earning 22 cents. This 22 cents gets added to Mr. Red, who now totals 33 cents. And Mr. Red, who started at 11 cents, has earned 4 cents (33% × .11) on his own, creating Mr. Purple.

Phew! The final value after 12 months is: $1 + 1 + .33 + .04$ or about 2.37. Take some time to really understand what's happening with this growth:

- Each color earns interest on itself and "hands it off" to another color. The newly-created money can earn money of its own, and on the cycle goes.

- I like to think of the original amount (Mr. Blue) as never changing. Mr. Blue shovels money to create Mr. Green, a steady 33 every 4 months since Mr. Blue does not change. In the diagram, Mr. Blue has a blue arrow showing how he feeds Mr. Green.

- Mr. Green just happens to create and feed Mr. Red (green arrow), but Mr. Blue isn't aware of this.

- As Mr. Green grows over time (being constantly fed by Mr. Blue), he contributes more and more to Mr. Red. Between months 4-8 Mr. Green gives 11 cents to Mr. Red. Between months 8-12 Mr. Green gives 22 cents to Mr. Red, since Mr. Green was at 66 cents during Month 8. If we expanded the chart, Mr. Green would give 33 cents to Mr. Red, since Mr. Green reached a full dollar by Month 12.

Make sense? It's tough at first — I even confused myself a bit while putting the charts together. But see that each "dollar" creates little helpers, who in turn create helpers, and so on.

We get a formula by using 3 periods in our growth equation:

$$growth = (1 + 100\%/3)^3 = 2.37037...$$

We earned $1.37, even better than the $1.25 we got last time!

7.6   Can We Get Infinite Money?

Why not take even shorter time periods? How about every month, day, hour, or even nanosecond? Will our returns skyrocket?

Our return gets better, but only to a point. Try using different numbers of n in our magic formula to see our total return:

n	$(1 + 1/n)^n$
1	2
2	2.25
3	2.37
5	2.488
10	2.5937
100	2.7048
1,000	2.7169
10,000	2.71814
100,000	2.718268
1,000,000	2.7182804

The numbers get bigger and converge around 2.718. Hey... wait a minute... that looks like e!

Yowza. In geeky math terms, e is defined to be the total growth when we continuously compound 100% return on smaller and smaller time periods:

$$growth = e = \lim_{n \to \infty} \left(1 + \frac{1}{n}\right)^n$$

This limit appears to converge, and there are proofs to that effect. But as you can see, as we take finer time periods the total return stays around 2.718.

7.7   But What Does It All Mean?

The number e (2.718...) represents the maximum compound rate of growth from a process that grows at 100% for one time period. Sure, you start out expecting to grow from 1 to 2. But with each tiny step forward you create a little "dividend" that starts growing on its own. When all is said and done, you end up with e (2.718...) at the end of 1 time period, not 2.

So, if we start with $1.00 and compound continuously at 100% return we get 1e. If we start with $2.00, we get 2e. If we start with $11.79, we get 11.79e.

e is like a speed limit (like c, the speed of light) saying how fast you can possibly grow using a continuous process.

7.8   What About Different Rates?

Good question. What if we are grow at 50% annually, instead of 100%? Can we still use e?

Let's see. The rate of 50% compound growth would look like this:

$$\lim_{n \to \infty} \left(1 + \frac{.50}{n}\right)^n$$

Hrm. What can we do here? Well, imagine we break it down into 50 chunks of 1% growth:

$$\left(1 + \frac{.50}{50}\right)^{50} = (1 + .01)^{50}$$

Sure, it's not infinity, but it's pretty granular. Now imagine we broke down our "regular" rate of 100% into chunks of 1% growth as well:

$$e \approx \left(1 + \frac{1.00}{100}\right)^{100} = (1 + .01)^{100}$$

Ah, something is emerging here. In our regular case, we have 100 cumulative changes of 1% each. In the 50% scenario, we have 50 cumulative changes of 1% each.

Different Growth Rates

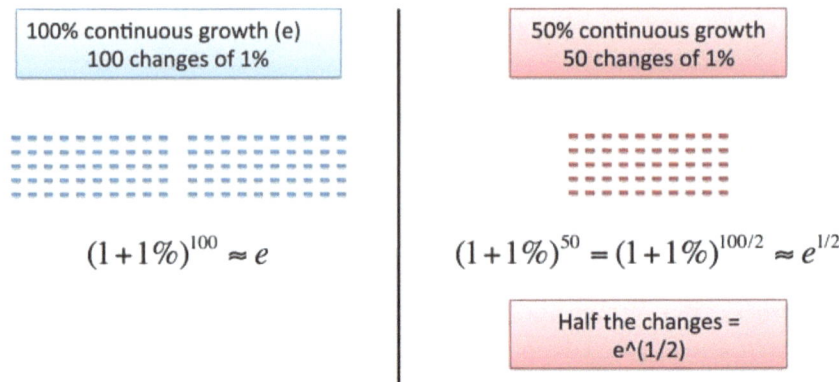

What is the difference between the two numbers? Well, it's just half the number of changes:

$$(1+.01)^{50} = (1+.01)^{100/2} = \left((1+.01)^{100}\right)^{1/2} = e^{1/2}$$

This is pretty interesting. 50 / 100 = .5, which is the exponent we raise e to. This works in general: if we had a 300% growth rate, we could break it into 300 chunks of 1% growth. This would be triple the normal amount for a net rate of e^3.

Even though growth can look like addition (+1%), we need to remember that it's really a multiplication (x 1.01). This is why we use exponents (repeated multiplication) and square roots ($e^{\frac{1}{2}}$ means "half" the number of changes, i.e. half the number of multiplications).

Although we picked 1%, we could have chosen any small unit of growth (.1%, .0001%, or even an infinitely small amount!). The key is that for any rate we pick, it's just a new exponent on e:

$$growth = e^{rate}$$

7.9 What About Different Times?

Suppose we have 300% growth for 2 years. We'd multiply one year's growth (e^3) by itself:

$$growth = \left(e^3\right)^2 = e^6$$

And in general:

$$growth = \left(e^{rate}\right)^{time} = e^{rate \cdot time}$$

Because of the magic of exponents, we can avoid having two powers and just multiply rate and time together in a single exponent.

7.10   The Big Secret: e Merges Rate and Time

This is wild! e^x can mean two things:

- x is the number of times we multiply a growth rate: 100% growth for 3 years is e^3

- x is the growth rate itself: 300% growth for one year is e^3.

Won't this overlap confuse things? Will our formulas break and the world come to an end?

It all works out. When we write:

$$e^x$$

the variable x is a combination of rate and time.

$$x = rate \cdot time$$

Let me explain. When dealing with compound growth, 10 years of 3% growth has the same overall impact as 1 year of 30% growth (and no growth afterward).

- 10 years of 3% growth means 30 changes of 1%. These changes happen over 10 years, so you are growing continuously at 3% per year.

- 1 period of 30% growth means 30 changes of 1%, but happening in a single year. So you grow for 30% a year and stop.

The same "30 changes of 1%" happen in each case. The faster your rate (30%) the less time you need to grow for the same effect (1 year). The slower your rate (3%) the longer you need to grow (10 years).

But in both cases, the growth is $e^{.30} = 1.35$ in the end. We're impatient and prefer large, fast growth to slow, long growth but e shows they have the same net effect.

So, our general formula becomes:

$$growth = e^x = e^{rt}$$

If we have a return of r for t time periods, our net compound growth is e^{rt}. This even works for negative and fractional returns, by the way.

7.11   Example Time!

Examples make everything more fun. A quick note: We're so used to formulas like 2^x and regular, compound interest that it's easy to get confused (myself included). Later in the book we'll cover simple, compound and continuous growth.

These examples focus on smooth, continuous growth, not the "jumpy" growth that happens at yearly intervals. There are ways to convert between them, which the interest chapter shows.

Example 1: Growing Crystals

Suppose I have 300kg of magic crystals. They're magic because they grow throughout the day: a single crystal, over 24 hours, smoothly sheds off its own weight in crystals. The baby crystals it made start growing immediately, but I can't track that – I'm watching how much the original sheds. How much will I have after 10 days?

First, this is a tricky example: we have an *input rate* of 100% growth every 24 hours, and want to know the compounded output rate after 10 days. The input rate is what the individual crystal knows about ("I need to generate my own weight over 24 hours") and the compounded output rate is what actually happens ("Those crystals you generated? Well they started growing on their own.").

e is the magic conversion factor: Our rate is 100% every 24 hours, so after 10 days we get: $300 \cdot e^{1 \cdot 10} = 6.6$ million kg of our magic gem.

In most circumstances, we don't know the internal rate of the phenomenon and first observe the final, compound output rate ($e^{100\%}$, i.e. seeing 1 crystal grow to 2.718 in 24 hours). Using the natural log we can deduce the initial rate of 100%.

Example 2: Maximum Interest Rates

Suppose I have \$120 in a count with 5% interest. My bank is generous and gives me the maximum possible compounding. How much will I have after 10 years?

Our rate is 5%, and we're lucky enough to compound continuously. After 10 years, we get $\$120 \cdot e^{.05 \cdot 10} = \$197.85$. Of course, most banks aren't nice enough to give you the best possible rate. The difference between your actual return and the continuous one is how much they don't like you.

Example 3: Radioactive Decay

I have 10kg of a radioactive material, which appears to *continuously* decay at a rate of 100% per year (that is, at the start of the year it appears to be shrinking at a rate of 10kg/year). How much will I have after 3 years?

Zip? Zero? Nothing? Think again.

Decaying continuously at 100% per year is the trajectory we *start off* with. Yes, we do begin with 10kg and expect to "lose it all" by the end of the year, since our initial rate is to decay 10 kg/year.

But riddle me this: we go a few months and are at 5kg. How much time is left: half a year, since we're losing 10kg/year?

Nope! Now that we have 5kg, we're only losing matter at a rate of 5kg/year, so we have a full year from this very moment!

Let's wait a few months and shrink to 2kg. See where this is going? Our rate is now to lose 2kg/year, so we have another year from this moment. We get to 1kg, have a full year, get to 0.5kg, have a full year — see the pattern?

As time goes on, we lose material, but our *rate of decay slows down*. This constantly slowing decay is the reverse of constantly compounding growth.

After 3 years, we'll have $10 \cdot e^{-1.3} = .498$ kg. We use a *negative exponent* for decay, which can be seen as:

- **Reversed growth (shrinking)**. A negative exponent gives us a fraction to shrink by $(1/e^{rt})$ vs a growth multiplier (e^{rt}).

- **Reversed time**. A negative exponent is like time going backwards; instead of seeing .498 grow to 10 (forward time), we start at 10 and go backwards to .498.

Negative exponential growth is just another way to change: you shrink, instead of grow.

More Examples

If you want fancier examples, try the Black-Scholes option formula (notice e used for exponential decay in value) or radioactive decay. The goal is to see e^{rt} in a formula and understand why it's there: it's modeling growth or decay.

And now you know why it's e, and not π or some other number: e raised to "$r \cdot t$" gives you the growth impact of rate r and time t.

7.12   There's More To Learn!

My goal was to:

- **Explain why e is important:** It's a fundamental constant, like pi, that shows up in growth rates.

- **Give an intuitive explanation:** e lets you see the impact of any growth rate. Every new "piece" (Mr. Green, Mr. Red, etc.) helps add to the total growth.

- **Show how it's used:** e^{rt} lets you predict the impact of any growth rate and time period.

- **Get you hungry for more:** In the upcoming chapters we'll dive into other properties of e.

This is just the start — cramming everything into a chapter would tire us both. Dust yourself off, take a break and learn about e's evil twin, the natural logarithm.

THE NATURAL LOGARITHM (LN)

The previous chapter was about understanding the exponential function; our next target is the natural logarithm.

Given how the natural log is described in math books, there's little "natural" about it: it's defined as the inverse of e^x, a strange enough exponent already.

But there's a fresh, intuitive explanation: **The natural log gives you the time needed to reach a certain level of growth.**

Suppose you have an investment in gummy bears (who doesn't?) with an interest rate of 100% per year, growing continuously. If you want 10x growth, assuming continuous compounding, you'd wait only $ln(10)$ or 2.302 years. Don't see why a few years of compounded growth can get 10x return? Read that chapter on e.

e and the natural log are twins:

- e^x is the amount of continuous growth after a certain amount of time

- The natural log (ln) is the amount of **time** needed to reach a certain level of continuous growth

Not too bad, right? While the mathematicians scramble to give you the long, technical explanation, let's dive into the intuitive one.

8.1 e is About Growth

The number e is about continuous growth. As we saw previously, e^x lets us merge rate and time: 3 years at 100% growth is the same as 1 year at 300% growth, when continuously compounded.

We can take any combination of rate and time (50% for 4 years) and convert the rate to 100% for convenience (giving us 100% for 2 years). By converting to a rate of 100%, we only have time to think about:

$$e^x = e^{rate \cdot time} = e^{1.0 \cdot time} = e^{time}$$

Intuitively, e^x means:

- How much growth do I get after after x units of time (and 100% continuous growth)

- For example: after 3 time periods I have $e^3 = 20.08$ times the amount of "stuff".

e^x is a scaling factor, showing us how much growth we'd get after x units of time.

8.2   Natural Log is About Time

The natural log is the inverse of e, a fancy term for opposite. Speaking of fancy, the Latin name is *logarithmus naturali*, giving the abbreviation *ln*.

Now what does this inverse or opposite stuff mean?

- e^x lets us plug in **time** and get growth.

- $ln(x)$ lets us plug in **growth** and get the **time it would take.**

For example:

- e^3 is 20.08. After 3 units of time, we end up with 20.08 times what we started with.

- $ln(20.08)$ is about 3. If we want growth of 20.08, we'd wait 3 units of time (again, assuming a 100% continuous growth rate).

With me? The natural log gives us the time needed to hit our desired growth.

8.3   Logarithmic Arithmetic Is Not Normal

You've studied logs before, and they were strange beasts. How'd they turn multiplication into addition? Division into subtraction? Let's see.

What is $ln(1)$? Intuitively, the question is: How long do I wait to get 1x my current amount?

Zero. Zip. Nada. You're already **at** 1x your current amount! It doesn't take any time to grow from 1 to 1.

$$ln(1) = 0$$

Ok, how about a fractional value? How long to get 1/2 my current amount? Assuming you are growing continuously at 100%, we know that $ln(2)$ is the amount of time to double. If we **reverse it** (i.e., take the negative time) we'd get half of our current value.

$$ln(.5) = -ln(2) = -.693$$

Makes sense, right? If we go backwards (negative time) .693 seconds we're at half our current amount. In general, you can flip the fraction and take the negative: $ln(1/3) = -ln(3) = -1.09$. This means if we go back 1.09 units of time, we'd have a third of what we have now.

Ok, how about the natural log of a negative number? How much time does it take to "grow" your bacteria colony from 1 to -3?

It's impossible! You can't have a "negative" amount of bacteria, can you? At most (er... least) you can have zero, but there's no way to have a negative amount of the little critters. Negative bacteria just doesn't make sense.

$$ln(negative\ number) = undefined$$

Undefined just means "there is no amount of time you can wait" to grow to a negative amount (we'll revisit this in Euler's Formula).

8.4 Logarithmic Multiplication is Mighty Fun

How long does it take to grow 4x your current amount? Sure, we could just use $ln(4)$. But that's too easy, let's be different.

We can consider 4x growth as doubling (taking $ln(2)$ units of time) and then doubling again (taking another ln(2) units of time):

- Time to grow 4x = $ln(4)$ = Time to double and double again = $ln(2) + ln(2)$

Interesting. Any growth number, like 20, can be considered 2x growth followed by 10x growth. Or 4x growth followed by 5x growth. Or 3x growth followed by 6.666x growth. See the pattern?

$$ln(a \cdot b) = ln(a) + ln(b)$$

The log of a times b = $log(a) + log(b)$. This relationship *makes sense* when you think in terms of time to grow.

If we want to grow 30x, we can wait $ln(30)$ all at once, or simply wait $ln(3)$, to triple, then $ln(10)$, to grow 10x again. The net effect is the same, so the net time should be the same too (and it is).

How about division? $ln(5/3)$ means: How long does it take to grow 5 times and then take 1/3 of that?

Well, growing 5 times is $ln(5)$. Growing 1/3 is $-ln(3)$ units of time. So

$$ln(5/3) = ln(5) - ln(3)$$

Which says: Grow 5 times and "go back in time" until you have a third of that amount, so you're left with 5/3 growth. In general we have

$$ln(a/b) = ln(a) - ln(b)$$

I hope the strange math of logarithms is starting to make sense: multiplication of growth becomes addition of time, division of growth becomes subtraction of time. Don't memorize the rules, **understand them**.

8.5 Using Natural Logs With Any Rate

"Sure," you say, "This log stuff works for 100% growth but what about the 5% I normally get?"

It's no problem. The "time" we get back from ln() is actually a combination of rate and time, the "x" from our e^x equation. We just assume 100% to make it simple, but we can use other numbers.

Suppose we want 30x growth: plug in $ln(30)$ and get 3.4. This means:

$$e^x = growth$$
$$e^{3.4} = 30$$

And intuitively this equation means "100% return for 3.4 years is 30x growth". We can consider the equation to be:

$$e^x = e^{rate \cdot time}$$
$$e^{100\% \cdot 3.4 years} = 30$$

We can modify "rate" and "time", as long as rate × time = 3.4. For example:

- 3.4 years at 100% = $3.4 \cdot 1.0 = 3.4$

- 1.7 years at 200% = $1.7 \cdot 2.0 = 3.4$

- 6.8 years at 50% = $6.8 \cdot 0.5 = 3.4$

- 68 years at 5% = $68 \cdot .05 = 3.4$

Cool, eh? The natural log can be used with any **interest rate or time** as long as their product is the same. You can wiggle the variables all you want.

8.6 Awesome Example: The Rule of 72

The Rule of 72 is a mental math shortcut to estimate the time needed to double your money. We're going to derive it (yay!) and even better, we're going to understand it intuitively.

How long does it take to double your money at 100% interest, compounded every year?

Uh oh. We've been using natural log for *continuous* rates, but now you're asking for *yearly* interest? Won't this mess up our formula? Yes, it will, but at reasonable interest rates like 5%, 6% or even 15%, there isn't much difference between yearly compounded and fully continuous interest. So the rough formula works, uh, roughly and we'll pretend we're getting fully continuous interest.

Now the question is easy: How long to double at 100% interest? $ln(2) = $ .693. It takes .693 units of time (years, in this case) to double your money with continuous compounding with a rate of 100%.

Ok, what if our interest isn't 100% What if it's 5% or 10%?

Simple. As long as rate × time = .693, we'll double our money:

$$rate \cdot time = .693$$
$$time = .693/rate$$

So, if we only had 10% growth, it'd take .693 / 10% or 6.93 years to double.

To simplify things, let's multiply by 100 so we can talk about 10 rather than .10:

- time to double = 69.3/rate, where rate is assumed to be in percent.

Now the time to double at 5% growth is 69.3/5 or 13.86 years. However, 69.3 isn't the most divisible number. Let's pick a close neighbor, 72, which can be divided by 2, 3, 4, 6, 8 and many more numbers.

- time to double = 72/rate

which is the rule of 72! Easy breezy.
If you want to find the time to triple, you'd use $ln(3) \sim 109.8$ and get

- time to triple = 110 / rate

Which is another useful rule of thumb. The Rule of 72 is useful for interest rates, population growth, bacteria cultures, and anything that grows exponentially.

8.7 Where To From Here?

I hope the natural log makes more sense — it tells you the **time** needed for any amount of exponential growth. I consider it "natural" because e is the universal rate of growth, so ln could be considered the "universal" way to figure out how long things take to grow.
When you see $ln(x)$, just think "the amount of time to grow to x".

8.8 Appendix: The Natural Log of e

Quick quiz: What's $ln(e)$?

- The math robot says: Because they are defined to be inverse functions, clearly $ln(e) = 1$

- The intuitive human: $ln(e)$ is the amount of time it takes to get "e" units of growth (about 2.718). But e **is** the amount of growth after **1 unit of time**, so $ln(e) = 1$.

Think intuitively.

INTEREST RATES

Interest rates are confusing, despite their ubiquity. This chapter takes an in-depth look at why interest rates behave as they do.

This concept will help you understand finance (mortgages & savings rates), along with the omnipresent e and natural logarithm. Here's our cheatsheet:

Term	Definition	Description & Usage
Simple Interest	$interest = P \cdot r \cdot n$	Fixed, non-growing return (bond coupons)
Compound Interest (Annual)	$total = P \cdot (1 + r)^n$	Changes each year (stock market, inflation)
Compound (t times per year)	$total = P \cdot (1 + r/t)^{tn}$	Changes each month/week/day (savings account)
Continuous Growth	$total = P \cdot e^{rt}$	Changes each instant (radioactive decay, temperature)
APR	Annual Percentage **Rate**	Nominal Return (compounding not included)
APY	Annual Percentage **Yield**	Actual Return (all compounding effects included)

9.1 Why the Fuss?

Interest rates are complex. Like Roman numerals and hieroglyphics, our first system "worked" but wasn't quite ideal.

In the beginning, you might have had 100 gold coins and were paid 12% per year (percent = per cent = per hundred — those Roman numerals still show up!). It's simple enough: we get 12 coins a year. But is it really 12?

If we break it down, it seems we earn 1 gold a month: 6 for January-June, and 6 for July-December. But wait a minute — after our June payout we'd have 106 gold in July, and yet earn only 6 during the rest of the year? Are you saying 100 and 106 earn the same amount in 6 months? By that logic, do 100 and 200 earn the same amount, too? Uh oh.

This issue didn't seem to bother the ancient Egyptians, but did raise questions in the 1600s and led to Bernoulli's discovery of e (sorry math fans, e wasn't discovered via some hunch that a strange limit would have useful properties). There's much to say about this riddle — just keep this in mind as we dissect interest rates:

- **Interest rates and terminology were invented before the idea of compounding.** Heck, loans were around in 1500 BC, before exponents, 0, or even the decimal point! So it's no wonder our discussions can get confusing.

- **Nature doesn't wait for a human year before changing.** Interest earnings are a type of "growth", but natural phenomena like temperature and radioactive decay change constantly, every second and faster. This is one reason why physics equations model change with "e" and not "$(1 + r)^n$": Nature rudely ignores our calendar when making adjustments.

9.2   Learn the Lingo

As a result of these complications, we need a few terms to discuss interest rates:

- **APR (annual percentage rate):** The rate someone tells you ("12% per year!"). You'll see this as "r" in the formula.

- **APY (annual percentage yield):** The rate you actually get after a year, after all compounding is taken into account. You can consider this "total return" in the formula. The APY is greater than or equal to the APR.

APR is what the bank tells you, the APY is what you pay (the price after taxes, shipping and handling, if you get my drift). And of course, banks advertise the rate that looks better.

Getting a credit card or car loan? They'll show the "low APR" you're paying, to hide the higher APY. But opening a savings account? Well, of course they'd tout the "high APY" they're paying to look generous.

The APY (actual yield) is what you care about, and the way to compare competing offers.

9.3   Simple Interest

Let's start on the ground floor: **Simple interest pays a fixed amount over time.** A few examples:

- Aesop's fable of the golden goose: every day it laid a single golden egg. It couldn't lay faster, and the eggs didn't grow into golden geese of their own.

- Corporate bonds: A bond with a face value of $1000 and 5% interest rate (coupon) pays you $50 per year until it expires. You can't increase the face value, so $50/year is what you will get from the bond.

Simple interest is the most **basic type of return**. Depositing $100 into an account with 50% simple (annual) interest looks like this:

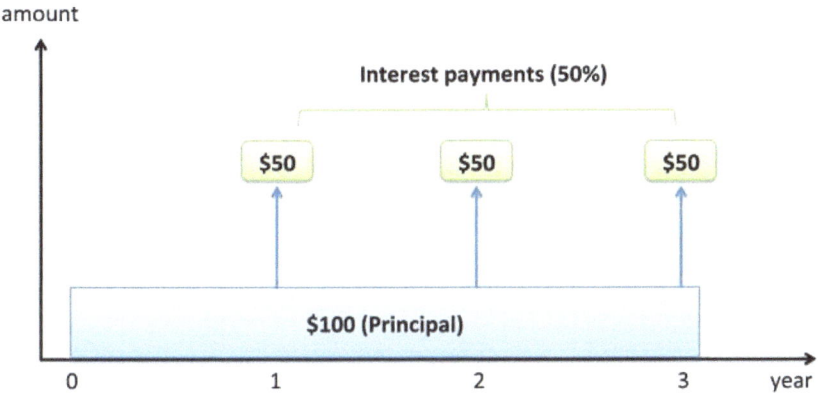

You start with a principal (aka investment) of $100 and earn $50 each year. I imagine the blue principal "shoveling" green money upwards every year.

However, this new, green money is stagnant — it can't grow! With simple interest, the $50 just sits there. Only the original $100 can do "work" to generate money.

Simple interest has a simple formula: Every period you earn P · r (principal · interest rate). After n periods you have:

$$interest = P \cdot r \cdot n$$

This formula works as long as "r" and "n" refer to the same time period. It could be years, months, or days — though in most cases, we're considering annual interest. There's no trickery because there's no compounding — interest can't grow.

Simple interest is useful when:

- **Your interest earnings create something that cannot grow more.** It's like the golden goose creating eggs, or a corporate bond paying money that cannot be reinvested.

- **You want simple, predictable, non-exponential results**. Suppose you're encouraging your kids to save. You could explain that you'll put aside $1/month in "fun money" for every $20 in their piggybank. Most kids would be thrilled and buy comic books each month. If your last name is Greenspan, your kid might ask to reinvest the dividend.

In practice, simple interest is fairly rare because most types of earnings can be reinvested. There really isn't an APR vs APY distinction, since your earnings can't change: you always earn the same amount per year.

9.4 Really Understanding Growth

Most interest explanations stop there: here's the formula, now get on your merry way. Not here: let's see what's really happening.

First, what does an interest rate mean? I think of it as a type of "speed":

- **50 mph** means you'll travel 50 miles in the course of an hour

- **r = 50% per year** means you'll earn 50% of your principal in the course of a year. If P = $100, you'll earn $50/year (your "speed of money growth").

But both types of speed have a subtlety: **we don't have to wait the full time period!**

Does driving 50 mph mean you must go a full hour? No way! You can drive "only" 30 minutes and go 25 miles (50 mph · .5 hours). You could drive 15 minutes and go 12.5 miles (50 mph · .25 hours). You get the idea.

Interest rates are similar. An interest rate gives you a "trajectory" or "pace" to follow. If you have $100 at a 50% simple interest rate, your pace is $50/year. But you don't need to follow that pace for a full year! If you grew for 6 months, you should be entitled to $25. Take a look at this:

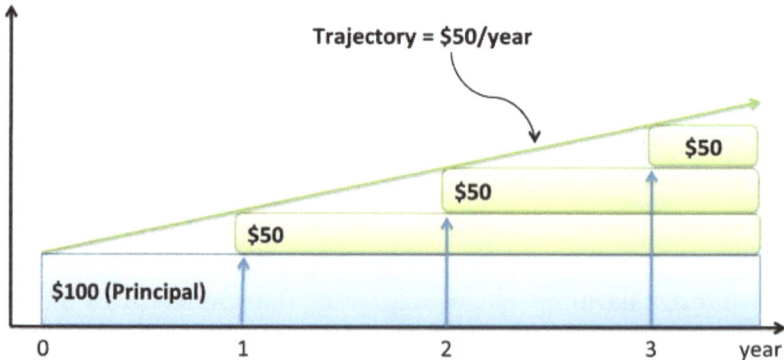

We start with $100, in blue. Each year that blue contributes $50 (in green) to our total amount. Of course, with simple interest our earnings are based on our original amount, not the "new total". Connecting the dots gives us a trendline: we're following a path of $50/year. Our payouts look like a staircase because we're only paid at the end of the year, but the trajectory still works.

Simple interest keeps the same trajectory: we earn "$P \cdot r$" each year, no matter what ($50/year in this case). That straight line perfectly predicts where we'll end up.

The idea of "following a trajectory" may seem strange, but stick with it — it will really help when understanding the nature of e.

One point: the trajectory is "how fast" a bank account is growing at a certain moment. With simple interest, we're stuck in a car going the same speed: $50/year, or 50 mph. In other cases, our rate may change, like a skydiver: they start off slow, but each second fall faster and faster. But at **any instant**, there's a single speed, a single trajectory.

(The math gurus will call this trajectory a "derivative" or "gradient". No need to hit a mosquito with the calculus sledgehammer just yet.)

9.5   Basic Compound Interest

Simple interest should make you squirm. **Why can't our interest earn money?** We should use the bond payouts ($50/year) to buy more bonds. Heck, we should use the golden eggs to fund research into cloning golden geese!

Compound growth means your interest earns interest. Einstein called it "one of the most powerful forces in nature", and it's true. When you have a growing thing, which creates more growing things, which creates more growing things. . . your return adds up fast.

The most basic type is period-over-period return, which usually means "year over year". Reinvesting our interest annually looks like this:

Compound Interest

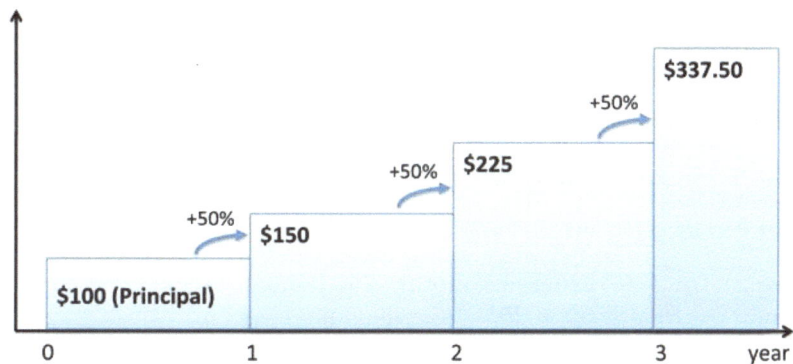

We earn $50 from year 0 – 1, just like with simple interest. But in year 1-2, now that our total is $150, we can earn $75 this year (50% · 150) giving us $225. In year 2-3 we have $225, so we earn 50% of that, or $112.50.

In general, we have $(1 + r)$ times more "stuff" each year. After n years, this becomes:

$$total = P \cdot (1+r)^n$$

Exponential growth outpaces simple, linear interest, which only had $250 in year 3 (100 + 3*50). Compound growth is useful when:

- **Interest can be reinvested**, which is the case for most savings accounts.

- **You want to predict a future value based on a growth trend.** Most trends, like inflation, GDP growth, etc. are assumed to be "compound-able". Yearly GDP growth of 3% over 10 years is really $(1.03)^{10} = 1.344$, or a 34.4% increase over that decade.

9.6 Interest as a Factory

The typical interpretation sees money as a "blob" that grows over time. This view works, but sometimes I like to see interest earnings as a "factory" that generates more money:

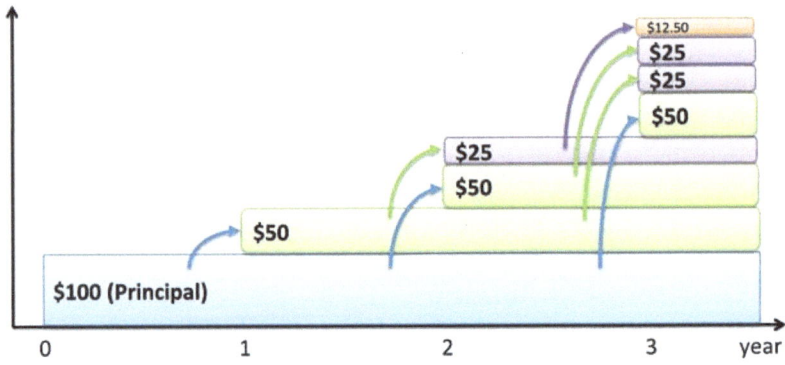

Here's what's happening:

- **Year 0:** We start with $100.

- **Year 1:** Our $100 creates a $50 "bond".

- **Year 2:** The $100 generates another $50 bond. The $50 generates a $25 bond. The total is 50 + 25 = 75, which matches up.

- **Year 3:** Things get a bit crazy. The $100 creates a third $50 bond. The two existing $50 bonds make $25 each. And the $25 makes a 12.50.

- **Years 4 to infinity:** Left as an exercise for the reader. (Don't you love that textbook cop out?)

This is an interesting viewpoint. The $100 just mindlessly cranks out $50 "factories", which start earning money independently (notice the 3 blue arrows from the blue principal to the green $50s). These $50 factories create $25 factories, and so on.

The pattern seems complex, but it's simpler in a way as well. The $100 has no idea what those zany $50s are up to: as far as the $100 knows, we're only making $50/year.

So why's this viewpoint useful?

- **You can separate the impact of the parent ($100) from the children.** For example, at Year 3 we have $328 total. The parent has earned $150 ("3 · 50% · $100 = $150", using the simple interest formula!). This means the "children" have contributed $328 − $150 − $100 = $128, or about 1/3 the total value.

- **Breaking earnings into components helps understand e.** Knowing more about e is a good thing because it shows up everywhere.

And besides, seeing old ideas in a new light is always fun. For one of us, at least.

9.7   Understanding the Trajectory

Oh, we're not done yet. One more insight — take a look at our trajectory:

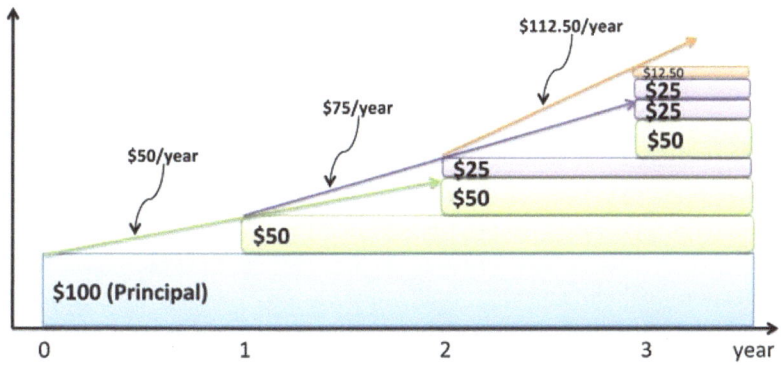

With simple interest, we kept the same pace forever ($50/year — pretty boring). With annually compounded interest, **we get a new trajectory each year**.

We deposit our money, go to sleep, and wake up at the end of the year:

- **Year 1:** "Hey, waittaminute. I've got $150 bucks! I should be making $75/year, not $50!". You yell at your banker, crank up the dial to $75/year, and go to sleep again.

- **Year 2:** "Hey! I've got $225, and should be making $112.50 per year!". You scream at your bank and get the rate adjusted.

This process repeats forever — we seem to never learn.

9.8 Compound Interest Revisited

Why are we waiting so long? Sure, waiting a year at a time is better than waiting "forever" (like simple interest), but I think we can do better. Let's zoom in on a year:

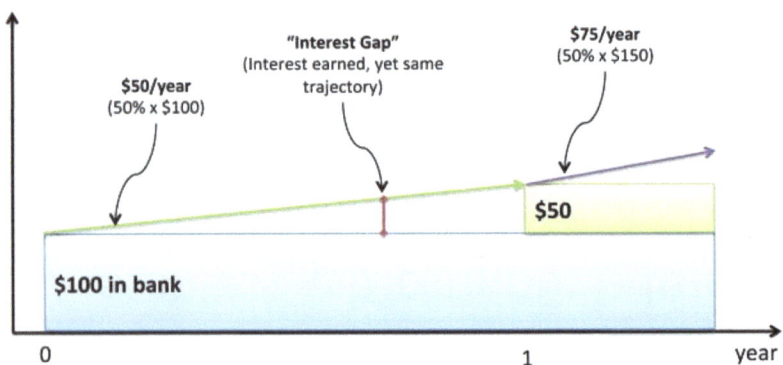

Look at what's happening. The green line represents our starting pace ($50/year), and the solid area shows the cash in our account. After 6 months, we've earned $25 but don't see a dime! More importantly, after 6 months we have the same trajectory as when we started. The **interest gap** shows where we've earned interest, but stay on our original trajectory (based on the original principal). We're losing out on what we should be making.

Imagine I took your money and returned it after 6 months. *"Well, ya see, I didn't use it for a full year, so I don't really owe you any interest. After all, interest is measured per year. Per yeeeeeaaaaar. Not per 6 months."* You'd smile and send Bubba to break my legs.

Annual payouts are man-made artifacts, used to keep things simple. But in reality, money should be earned all the time. We can pay interest after 6 months to reduce the gap:

Twice-Compounded Interest

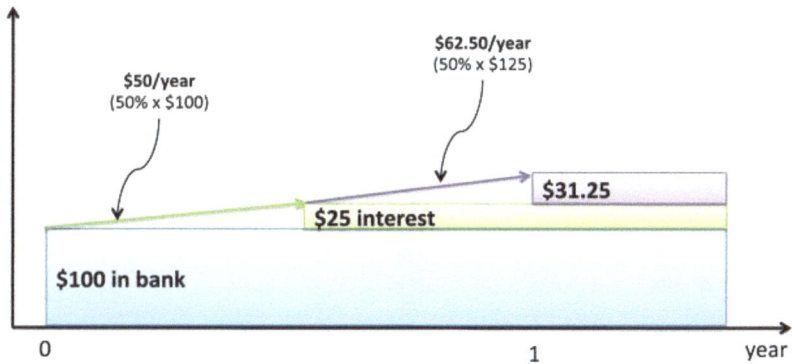

Here's what happened:

- We start with $100 and a trajectory of $50/year, like normal

- After 6 months we get $25, giving us $125

- We head out using the new trajectory: $50\% \cdot \$125 = \$62.5$/year

- After 6 months we collect 62.5/year times .5 year = 31.25. We have 125 + 31.25 = 156.25.

The key point is that our trajectory improved halfway through, and we earned 156.25, instead of the "expected" 150. Also, early payout gave us a smaller gap area (in white), since our $25 of interest was doing work for the second half (it contributed the extra 6.25, or $25 \cdot 50\% \cdot .5$ years).

For 1 year, the impact of rate r compounded t times is:

$$(1 + r/t)^t$$

In our case, we had $(1 + 50\%/2)^2$. Repeating this for n years (multiplying n times) gives:

$$total = P \cdot (1 + r/t)^{tn}$$

Compound interest reduces the "dead space" where our interest isn't earning interest. The more frequently we compound, the smaller the gap between earning interest and updating the trajectory.

9.9 Continuous Growth

Clearly we want money to "come online" as fast as possible. Continuous growth is compound interest on steroids: you shrink the gap into oblivion, by dividing the year into more and more time periods:

Continuous Growth

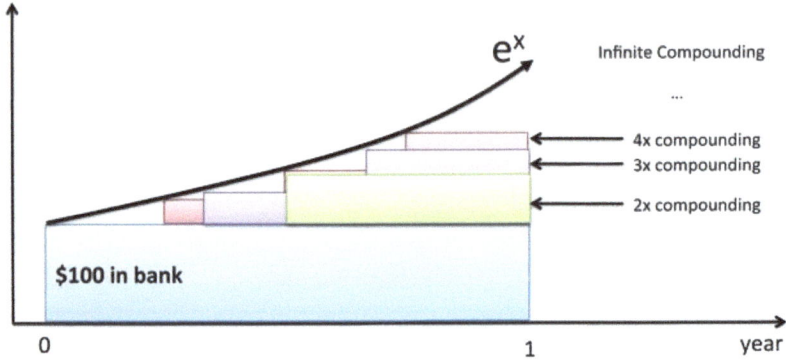

The net effect is to make use of interest as soon as it's created. We wait a millisecond, find our new sum, and go off in the new trajectory. Except it's not every millisecond: it's every nanosecond, picosecond, femtosecond, and intervals I don't know the name for. Continuous growth keeps the trajectory perfectly in sync with your current amount.

Review the chapter on e for more details. If we have rate r and time t (in years), the result is:

$$total = P \cdot e^{rt}$$

If you have a 50% APR, it would be an APY of $e^{.50} = 64.9\%$ if compounded continuously. That's a pretty big difference! Notice that e takes care of the icky parts, like dividing by an infinite number of periods.

Why's this useful?

- **Most natural phenomena grow continuously.** As mentioned earlier, physical phenomena grow on their own schedule: radioactive material doesn't wait for the Earth to go around the Sun before deciding to decay. Any physical equation that models change is going to use e^{rt}.

- **e^{rt} is the adjustable, one-size-fits-all exponential.** It sounds strange, but e can even model the jumpy, staircase-like growth we've seen with compound interest.

Most interest discussions leave e out, as continuous interest is not often used in financial calculations. (Daily compounding, $(1 + r/365)^{365}$, is generous enough for your bank account, thank you very much. But seriously, daily compounding is a pretty good approximation of continuous growth.)

The exponential e is the bridge from our jumpy "delayed" growth to the smooth changes of the natural world.

9.10   A Few Examples

Let's try a few examples to make sure it's sunk in. Remember: the APR is the rate they give you, the APY is what you actually earn (your true return).

- **Is a 4.5 APY better than a 4.4 APR, compounded quarterly?** You need to compare APY to APY. 4.4% compounded quarterly is $(1 + 4.4\%/4)^4 = 4.47\%$, so the 4.5% APY is still better.

- **Should I pay my mortgage at the end of the month, or the beginning?** The beginning, for sure. This way you knock out a chunk of debt early, preventing that "debt factory" from earning interest for 30 days. Suppose your loan APY is 6% and your monthly payment is $2000. By paying at the start of the month, you'd save $2000 \cdot 6\%$ = $120/year, or $3600 throughout a 30-year mortgage. And a few grand is nothing to sneeze at.

- **Should I use several small payments, or one large payment?.** You want to pay debt off as early as possible. $500/week for 4 weeks is better than $2000 at the end of the month. Each payment stops a few weeks' worth of interest. The math is a bit tricker, but think of it as 4 $500 investments, each getting a different return. In a month, the first payment saves 3 weeks worth of interest: $500 \cdot (1 + \text{daily rate})^{21}$. The next saves 2 weeks: $500 \cdot (1 + \text{daily rate})^{14}$. The third saves a week $500 \cdot (1 + \text{daily rate})^{7}$ and the last payment doesn't save any interest. Regardless of the details, **prepayment will save you money.**

The general principle: When investing, get interest paid early, so it can compound. When borrowing, pay debt early to *prevent* that interest from compounding.

9.11   Onward and Upward

This is a lot for one sitting, but I hope you've seen the big picture:

- **The interest rate (APR) is the "speed" at which money grows**.

- **Compounding lets you adjust your "speed" as you earn more interest.** The APR is the initial speed; the APY is the actual change during the year.

- **Man-made growth uses** $(1 + r)^n$, or some variant. We like our loans to line up with years.

- **Nature uses** e^{rt}. The universe doesn't particularly care for our solar calendar.

- **Interest rates are tricky.** When in doubt, ask for the APY and pay debt early.

Treating interest in this funky way (trajectories and factories) will help us understand some of e's cooler properties, which come in handy for calculus. Also, try the Rule of 72 for a quick way to compute the effect of interest rates mentally (that investment with 6% APY will double in 12 years). Happy math.

10

UNDERSTANDING EXPONENTS

We're taught that exponents are repeated multiplication. This is a good introduction, but it breaks down on $3^{1.5}$ and the brain-twisting 0^0. How do you repeat zero "zero" times and get 1 — without melting your brain?

You can't, not while exponents are repeated multiplication. Today our mental model is due for an upgrade.

Updated Mental Models

Operation	Old Concept	New Concept
Addition	Repeated counting	Sliding
Multiplication	Repeated addition	Scaling
Exponents	Repeated multiplication	Growing for amount of time

10.1   Viewing Arithmetic As Transformations

Let's step back — how do we learn arithmetic? We're taught that numbers are counts of something (fingers), addition is combining those counts ($3 + 4 = 7$) and multiplication is repeated addition ($2 \times 3 = 2 + 2 + 2 = 6$).

This interpretation works for round numbers like 2 and 10. Strange concepts like -1 and $\sqrt{2}$ seem to fit. Why?

Our model wasn't complete. Numbers aren't just a count; a better viewpoint is a *position on a line*. This position can be negative (-1), between other numbers ($\sqrt{2}$), or in another dimension (i).

Arithmetic became a general way to transform a number. Addition is sliding along the number line (+3 means slide 3 to the right) and multiplication is scaling (×3 means scale it up 3x).

So what are exponents?

10.2   Enter the Expand-o-tron

Let me introduce the Expand-o-tron 3000.

Expand-o-Tron 3000

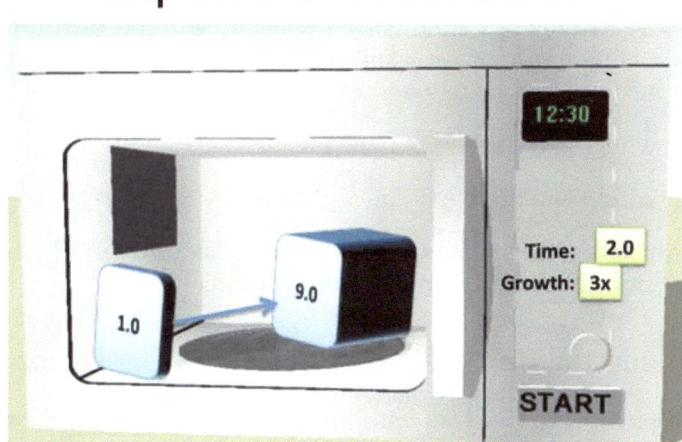

Yes, this device *looks* like a shoddy microwave — but instead of heating food, it grows numbers. Put a number in and a new one comes out. Here's how:

- Start with 1.0

- Set the *growth* to the desired change after one second (2x, 3x, 10.3x)

- Set the *time* to the number of seconds to grow

- Push start

And shazam! The bell rings and we pull out our shiny new number. Suppose we want to change 1.0 into 9:

- Put 1.0 in the expand-o-tron

- Set the change for "3x" growth, and the time for 2 seconds

- Push start

The number starts transforming as soon as we begin: We see 1.0, 1.1, 1.2...and just as we finish the first second, we're at 3.0. And the growth continues: 3.1, 3.5, 4.0, 6.0, 7.5...And just as the 2nd second ends we're at 9.0. Behold our shiny new number!

Mathematically, the expand-o-tron (exponent function) does this:

$$original \cdot growth^{duration} = new$$

or

$$growth^{duration} = \frac{new}{original}$$

For example, $3^2 = 9/1$. The base is the amount to grow each unit (3x), and the exponent is the amount of time (2). A formula like 2^n means "Use the expand-o-tron at 2x growth for n seconds".

Remember, we always start with 1.0 in the expand-o-tron to see how it changes a single unit. If we want to see what would happen if we started with 3.0 in the expand-o-tron, we just scale up the final result. For example:

- "Start with 1 and double 3 times" means $1 \cdot 2^3 = 1 \cdot 2 \cdot 2 \cdot 2 = 8$

- "Start with 3 and double 3 times" means $3 \cdot 2^3 = 3 \cdot 2 \cdot 2 \cdot 2 = 24$

Whenever you see a plain exponent (like 2^3), we're implicitly starting with 1.0 and transforming with 2x growth for 3 seconds.

10.3 Understanding the Exponential Scaling Factor

When multiplying, we can just state the final scaling factor. Want it 8 times larger? Multiply by 8. Done.

Exponents are a bit. . . finicky:

You: I'd like to grow this number.
Expand-o-tron: Ok, stick it in.
You: How big will it get?
Expand-o-tron: Gee, I dunno. Let's find out. . .
You: Find out? I was hoping you'd kn-
Expand-o-tron: Shh!!! It's growing! It's growing!
You: . . .
Expand-o-tron: It's done! My masterpiece is alive!
You: Can I go now?

The expand-o-tron is indirect. Just looking at it, you're not sure what it'll do: What does 3^{10} mean to you? How does it make you feel? Instead of a nice finished scaling factor, exponents want us to feel, relive, even smell the growing process. Whatever you end with is your scaling factor.

It sounds roundabout and annoying. You know why? **Most things in nature don't know where they'll end up!**

Do you think bacteria *plans* on doubling every 14 hours? No — it just eats the moldy bread you forgot about in the fridge as fast as it can, and as it gets bigger the blob starts growing even faster (a purely hypothetical situation, of course). To predict the behavior, we input how fast they're growing (current rate) and how long they'll be changing (time) to work out their final value.

The answer has to be worked out — exponents are a way of saying "Begin with these conditions, start changing, and see where you end up". The expand-o-tron (or our calculator) does the work by crunching the numbers to get the final scaling factor. But someone has to do it.

10.4 Understanding Fractional Powers

Let's see if the expand-o-tron can help us understand exponents. First up: what does $2^{1.5}$ mean?

It's confusing when we think of repeated multiplication. But the expand-o-tron makes it simple: 1.5 is just the amount of time in the machine.

- 2^1 means 1 second in the machine (2x growth)

- 2^2 means 2 seconds in the machine (4x growth)

So $2^{1.5}$ means 1.5 seconds in the machine, so somewhere between 2x and 4x growth. The idea of "repeated counting" had us stuck with integers.

10.5 Multiplying Exponents

What if we want two growth cycles back-to-back? Let's say we use the machine for 2 seconds, and then use it for 3 seconds at the exact same power:

$$x^2 \cdot x^3 = ?$$

Think about your regular microwave — isn't this the same as one continuous cycle of 5 seconds? It sure is. As long as the power setting (base) stayed the same, we can just add the time:

$$x^y \cdot x^z = x^{y+z}$$

Again, the expand-o-tron gives us a *scaling factor* to change our number. To get the total effect from two consecutive uses, we just multiply the scaling factors together.

10.6 Square Roots

Let's keep going. Let's say we're at power level a and grow for 3 seconds:

$$a^3$$

Not too bad. Now what would growing for half that time look like? It'd be 1.5 seconds:

$$a^{1.5}$$

Now what would happen if we did that twice?

$$a^{1.5} \cdot a^{1.5} = a^3$$

Said another way: partial growth × partial growth = full growth

Looking at this equation, we see "partial growth" is the square root of full growth! If we divide the *time* in half we get the *square root* scaling factor. And if we divide the time in thirds?

$$a^1 \cdot a^1 \cdot a^1 = a^3$$

Or: partial growth × partial growth × partial growth = full growth

And we get the cube root! For me, this is an *intuitive* reason why dividing the exponents gives roots: we split the time into equal amounts, so each "partial growth" period must have the same effect. If three identical effects are multiplied together, it means they're each a cube root.

10.7 Negative Exponents

Now we're on a roll — what does a negative exponent mean? Well, "negative seconds" means going back in time! If going forward grows by a scaling factor, going backwards should shrink by one.

$$2^{-1} = \frac{1}{2^1}$$

The sentence means "1 second ago, we were at half our current amount (aka $1/2^1$)". In fact, this is a neat part of any exponential graph, like 2^x:

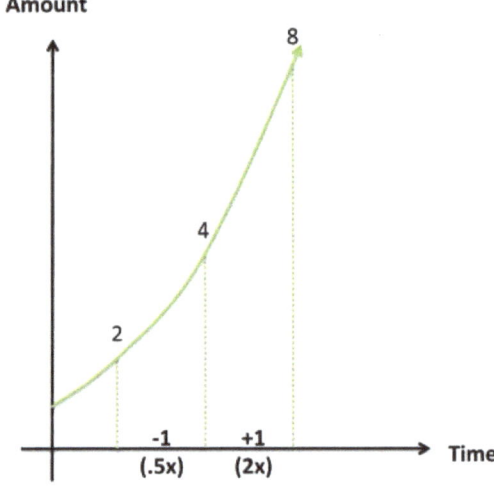

Pick a point like 3.5 seconds ($2^{3.5} = 11.3$). One second in the future we'll be at double our current amount ($2^{4.5} = 22.5$). One second ago we were at half our amount ($2^{2.5} = 5.65$).

This works for any number! Wherever 1 million is in our doubling growth curve, we were at 500,000 one second before it.

10.8 Taking the Zeroth Power

Now let's try the tricky stuff: what does 3^0 mean? Well, we set the machine for 3x growth, and use it for... *zero seconds*. Zero seconds means we don't even use the machine!

Our new and old values are the same (new = old), so the scaling factor is 1. Using 0 as the time (power) means there's no change at all. The scaling factor is always 1.

10.9 Taking Zero As a Base

How do we interpret 0^x? Well, our growth amount is "0x" — after a second, the expand-o-tron obliterates the number and turns it to zero. But if we've obliterated the number after 1 second, it really means any amount of time will destroy the number:

$$0^{1/n} = \text{nth root of } 0^1 = \text{nth root of } 0 = 0$$

No matter the tiny power we raise it to, it will be *some* root of 0.

10.10 Zero to The Zeroth Power

At last, the dreaded 0^0. What does it mean? The expand-o-tron to the rescue:

$$0^0 \text{ means a 0x growth for 0 seconds!}$$

Although we *planned* on obliterating the number, we never used the machine. No usage means new = old, and the scaling factor is just 1. $0^0 = 1 \cdot 0^0 = 1 \cdot 1 = 1$ — it doesn't change our original number. Mystery solved!

(For the math geeks: Defining 0^0 as 1 makes many theorems work smoothly. In reality, 0^0 depends on the scenario (continuous or discrete) and is under debate. The microwave analogy isn't about rigor: it helps us see why $0^0 = 1$ can be reasonable, in a way that "repeated counting" does not.)

10.11 Advanced: Repeated Exponents (a to the b to the c)

Repeated exponents are tricky. What does this mean?

$$(2^a)^b$$

It's "repeated multiplication, repeated" — another way of saying "do that exponent thing once, and do it again". Let's dissect it:

$$(2^3)^4$$

- First, I want to grow by doubling each second: do that for 3 seconds (2^3)

- Then, whatever my number is (8x), I want to grow by *that new amount* for 4 seconds (8^4)

The first exponent (^{3}) just knows to take "2" and grow it by itself 3 times. The next exponent (^{4}) just knows to take the previous amount (8) and grow it by itself 4 times. Each time unit in "Phase II" is the same as repeating all of Phase I:

$$(2^3)^4 = 2^3 \times 2^3 \times 2^3 \times 2^3 = 2^{3+3+3+3} = 2^{12}$$

Repeated counting helps us get our bearings. But then we bring out the expand-o-tron analogy: we grow for 3 seconds in Phase I, and redo that for 4 more seconds in Phase II. The expand-o-tron works for fractional powers:

$$(2^{3.1})^{4.2}$$

which means "Grow for 3.1 seconds, and use that new growth rate for 4.2 seconds". We can smush together the time (3.1 × 4.2) like this:

$$(a^b)^c = a^{b \cdot c} = (a^c)^b$$

Repeated exponents is a bit strange, so try some examples:

- $(2^1)^x$ means "Grow at 2 for 1 second, and 'do that growth' for x more seconds".

- $7 = (7^{0.5})^2$ means "We can jump to 7 all at once. Or, we can plan on growing to 7 but only use half the time ($\sqrt{7}$). But we can do that process for 2 seconds, which gives us the full amount ($\sqrt{7}$ squared = 7)."

We're like kids learning that $3 \times 7 = 7 \times 3$.

10.12 Advanced: Rewriting Exponents For The Grower

The expand-o-tron is a bit strange: numbers start growing the instant they're inside, but we specify the desired growth at the *end* of each second.

We say we want 2x growth at the *end* of the first second. But how do we know what rate to start off with? How far along should we be at 0.5 seconds? It can't be the full amount, or else we'll overshoot our goal as our interest compounds.

Here's the key: **Growth curves written like 2^x are from the observer's viewpoint, not the grower.**

The value "2" is measured at the *end* of the interval and we work backwards to create the exponent (Oh, it looks like you're growing at 2^x). This is convenient for us, but not the growing quantity — bacteria, radioactive elements and money don't care about lining up with our ending intervals!

No, these critters know their *current, instantaneous growth rate*, and don't try to line up their final amounts with our boundaries. It's just like understanding radians vs. degrees — radians are "natural" because they are measured from the mover's viewpoint.

To get into the grower's viewpoint, we use the magical number e. There's much more to say, but we can convert any "observer-focused" formula like 2^x into a "grower-focused" one:

$$2^x = (e^{ln(2)})^x = e^{ln(2)x}$$

In this case, $ln(2) = .693 = 69.3\%$ is the instantaneous growth rate needed to look like 2^x to an observer. When you ask for "2x growth at the end of each period", the expand-o-tron knows this means to grow the number at a rate of 69.3%.

There's more details, but remember this:

- The instantaneous growth rate controlled by the bacteria

- The overall rate measured at the end of each interval by the observer

Underneath it all, every exponential curve is just a scaled version of e^x:

$$a^x = (e^{ln(a)})^x = e^{ln(a)x}$$

Every exponent is a variation of e, just like every number is a scaled version of 1.

10.13   Why Use This Analogy?

Does the expand-o-tron exist? Do numbers really gather up in a number line? Nope — they're ways of looking at the world.

The expand-o-tron removes the mental hiccups when seeing $2^{1.5}$ or even 0^0. Everything from slide rules to Euler's Formula begins to click once we recognize the core theme of growth — even beasts like i^i can be tamed.

Friends don't let friends think of exponents as repeated multiplication. Happy math.

11

EULER'S FORMULA

Euler's formula looks utterly baffling:

$$e^{ix} = cos(x) + i\,sin(x)$$

This means

$$e^{i\pi} = cos(\pi) + i\,sin(\pi) = -1 + i(0) = -1$$

which is so surreal I need to write it again:

$$e^{i\pi} = -1$$

The equation relates an *imaginary exponent* to sine and cosine. And somehow you put in an infinite decimal like π and it nonchalantly pops out -1? Could this ever have an intuitive meaning?

Not according to 1800s mathematician Benjamin Peirce: "It is absolutely paradoxical; we cannot understand it, and we don't know what it means, but we have proved it, and therefore we know it must be the truth."

This attitude makes my blood boil. Should we throw up our hands and memorize? No!

Euler's formula describes two ways to move around a circle. That's it? One of the most stunning equations is just about spinning around? You got it – and today we'll understand why.

11.1 Understanding $cos(x) + i\,sin(x)$

The equal sign is overloaded. Sometimes it means "set one thing to another" (like $x = 3$) and other times it means "these are two ways of describing the same thing" (like $\sqrt{-1} = i$).

Euler's formula is equating two ways to describe the same movement: traveling around a circle. For our purposes, if you travel a distance of x radians:

Traversing A Circle

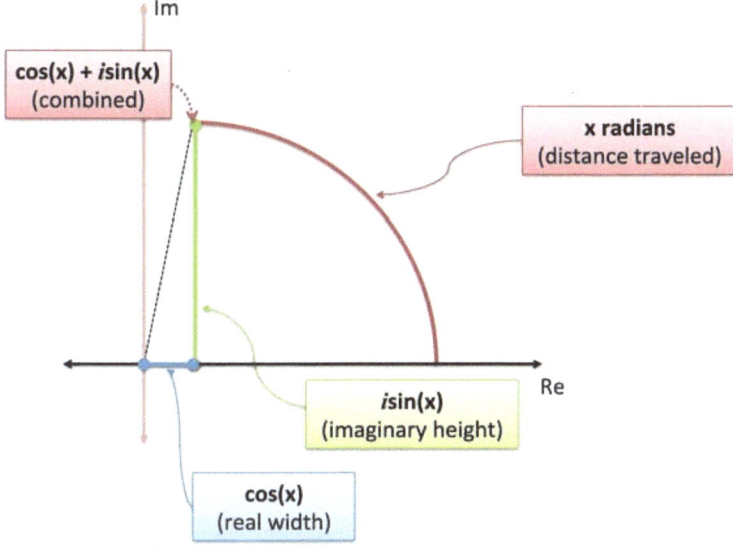

- $cos(x)$ is the x-coordinate (horizontal distance)

- $sin(x)$ is the y-coordinate (vertical distance)

$cos(x) + isin(x)$ is a clever way to put both coordinates in a single complex number. The "complex numbers are 2-dimensional" analogy helps interpret this as the position on a 2-d circle. Remember our circle definitions in the first chapter? Now we have a new one to add.

When we plug in $x = \pi$ (*setting* x to be π, just for this case) it means we're traveling along the outside of the unit circle. Since the total circumference is 2π, regular π is halfway around.

Going π radians takes us from 1.0, our starting point on the unit circle, all the way around to -1.0. There's no imaginary part (y-coordinate) because -1 fits firmly on the regular number line. And if we used $x = -\pi$ we'd travel clockwise through the bottom and get the same result: -1.

Cool. So Euler's formula says that e^{ix} is the same process as moving around a circle using $(cos(x) + isin(x))$. Now let's figure out how the e side of the equation accomplishes it.

11.2   What is Imaginary Growth?

Seeing imaginary numbers describe x- and y- coordinates is slightly tricky, but manageable. But what does an imaginary exponent mean?

Let's get back to the expand-o-tron. Something like 3^4 means "Have 3x growth for 4 seconds". And from the point of view of the *grower*:

$$3^4 = (e^{ln(3)})^4 = e^{ln(3)\cdot 4}$$

The grower just knows its instantaneous rate ($ln(3)$, which works out to 3 after all the compounding) and it wants us to *transform* that rate by 4. So it scales the rate 4x and off it goes: $e^{ln(3)\cdot 4} = 81$.

Now, why did it scale the rate 4 times? Because that's what multiplying by 4 (a real number) does. But imaginary numbers are different: when they're used (multiplied), they rotate the result.

11.3 Rotated Growth

Regular growth "pushes" a number in the same direction: 2×3 pushes 2 in its original direction, making it 3 times larger (6).

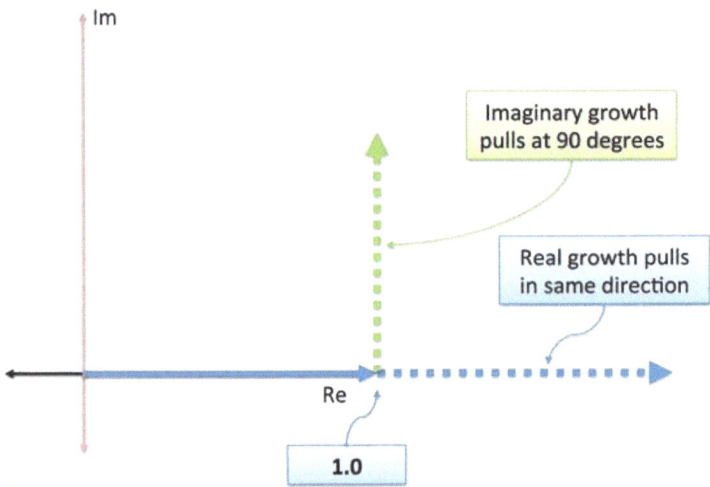

But an imaginary growth rate should give you "interest" in the imaginary direction and it would start pulling you 90-degrees! The neat thing about an orthogonal (perpendicular) push is that it doesn't speed you up or slow you down – it rotates you! Taking any number and multiplying by i will not change its magnitude, just the direction it points.

Intuitively, when we take an *imaginary growth rate* what we're really saying is:

- **Imaginary growth**: When I grow, don't push me forward or back in the direction I'm already going. Rotate me instead.

A constant rate of rotation will not change your size – you'll just be spinning around in a circle!

11.4   But Shouldn't We Spin Faster and Faster?

Nope. Here's why: regular growth keeps pushing you along in your original direction. So you go 1, 2, 4, 8, 16, multiplying 2x each time and staying in the real numbers.

But purely imaginary growth keeps you rotating. Let's say your growth rate is 100% in the i direction: As you constantly push, you keep changing direction and only get the benefit of rotation.

After 1 second you'll be at 90 degrees (i), at two seconds 180 degrees ($i^2 = -1$), and so on. Imaginary growth doesn't compound! If your growth rate is a larger imaginary number (2i), you can consider the growth happening for twice as long (remember how e merges rate and time?). But it's still pushing you in a perpendicular direction, which doesn't change your speed.

Now, if your growth rate is complex (a+bi) then the real part (a) grows or shrinks your magnitude like a normal exponent does, while the imaginary part (bi) rotates you. But Euler's formula (as written) is about purely imaginary growth (e^{ix}). We'll get into complex growth in a bit.

11.5   The Nitty Gritty Details

Let's take a closer look. Remember this definition of e:

$$e = \lim_{n \to \infty} \left(1 + \frac{100\%}{n}\right)^n$$

That $\frac{1}{n}$ represents the interest we earned in our period. We assumed the interest was real – but what if it was imaginary?

$$e = \lim_{n \to \infty} \left(1 + \frac{100\% \cdot i}{n}\right)^n$$

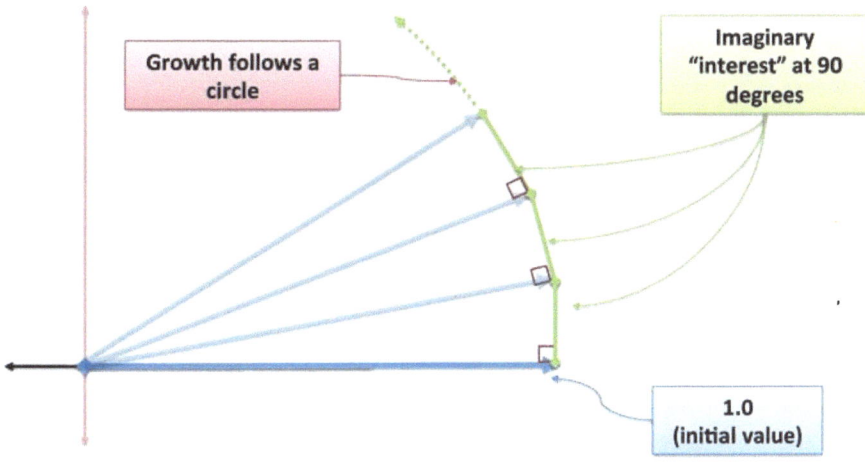

Cumulative Imaginary Interest

Now our interest pushes us in the 90-degree direction, which does not affect our length. (This is a tricky concept, because it seems we make a triangle where the hypotenuse must be larger. We are dealing with a limit; the extra distance from the hypotenuse is not detectable by us in any error margin we specify. We'll need calculus to help sort this out, but this is for another day).

We're applying i units of growth in infinitely small amounts. Each part is nudging us along at a 90-degree angle. There is no "faster and faster" rotation because the growth is always orthogonal, pushing us in a new direction ($+1$ degree).

So, we've found another way to describe a circle!

- **Circular motion**: Change continuously by rotating at 90-degree angle (imaginary growth rate).

So, Euler's formula is saying that "exponential imaginary growth traces out a circle". And this tracing is the same as describing circular growth using sine and cosine on the imaginary place. In this case "exponential" is a bit of misnomer since we move around the circle at a constant rate (a better word may be "continuously change"). But the majority of the time we're dealing with real exponents which do have a cumulative, compounding effect.

11.6   Some Examples

You don't really believe me. Here's some examples and how to intuitively think about them.

Example: e^i

Where's the x? Ah, it's just 1. Intuitively, without breaking out a calculator, we know that this means "travel 1 radian along the unit circle":

$$e^i = cos(1) + isin(1) = .5403 + .8415i$$

Not the prettiest number, but it works. Remember to put your calculator in radian mode when punching this in.

Example: 3^i

This is tricky – it's not in our standard format. But remember, $3^i = 1 \cdot 3^i$ – the real question is "How do we transform 1"?

We want an initial growth of 3x at the end of the period, or an instantaneous rate of ln(3). But, the i comes along and changes that rate of $ln(3)$ to $ln(3) \cdot i$:

$$3^i = (e^{ln(3)})^i = e^{ln(3) \cdot i}$$

We *thought* we were going to transform at a regular rate of ln(3) (a little faster than 100% since e is about 2.718). But oh no, i spun us around: now we're transforming at an imaginary rate which means we're just rotating about.

If i was a regular number like 4, it would have made us grow 4x faster. Now we're growing at a speed of ln(3), but sideways.

We should expect a complex number on the unit circle – there's nothing in the growth rate to increase our size. Solving the equation:

$$3^i = e^{ln(3) \cdot i} = cos(ln(3)) + i \, sin(ln(3)) = .4548 + .8906i$$

Example: i^i

In the past this would have sent me running, possibly in tears. But we can break it down into its transformations: $i^i = 1 \cdot i^i$. We start with 1 and want to transform it. Like solving 3^i, what's the instantaneous growth rate represented by i as a base?

Hrm. Normally we'd do $ln(x)$ to get the growth rate needed to reach x in one unit of time. But for an imaginary rate? We need to noodle this over.

In order to start with 1 and grow to i we need to start rotating. How fast? Well, we need to get 90 degrees ($\frac{\pi}{2}$ radians) in 1 unit of time. So our rate is $\frac{\pi}{2}i$ (Remember our rate must be imaginary since we're rotating! Plain old $\frac{\pi}{2} \sim 1.57$ results in regular growth.).

This should make sense: to turn 1.0 to i at the end of 1 unit, we should rotate $\frac{\pi}{2}$ radians (90 degrees) in that amount of time.

Phew. That describes the base. How about the exponent?

Well, the *other* i tells us to change our rate (yes, that rate we spent so long figuring out)! So rather than rotating at a speed of $\frac{\pi}{2}i$, which is what a base of i means, we transform the rate to:

$$\frac{\pi}{2}i \cdot i = \frac{\pi}{2} \cdot -1 = -\frac{\pi}{2}$$

The i's cancel and make the growth rate real again! We rotated our rate and pushed ourselves into the negative numbers. And a negative growth rate means we're shrinking – we should expect i^i to make things smaller. And it does: $i^i = e^{-\frac{\pi}{2}} \sim .2$. Tada! (Search "i^i=" on Google to use its calculator)

Take a breather: You can intuitively figure out how imaginary bases and imaginary exponents should behave. Whoa.

Example: $(i^i)^i$

More? If you insist. First off, we know what our growth rate will be inside the parenthesis:

$$i^i = (e^{\frac{\pi}{2}i})^i = e^{-\frac{\pi}{2}}$$

We get a negative (shrinking) growth rate of $\frac{\pi}{2}$. And now we modify that rate again by i:

$$(i^i)^i = (e^{-\frac{\pi}{2}})^i = e^{-\frac{\pi}{2}i}$$

And now we have a negative rotation! We're going around the circle a rate of $-\frac{\pi}{2}$ per unit time. How long do we go for? Well, there's an implicit "1" unit of time at the very top of this exponent chain; the implied default is to go for 1 time unit (just like $e = e^1$). 1 time unit gives us a rotation of $-\frac{\pi}{2}$ (-90 degrees) or $-i$!

$$i^i = .2078...$$
$$(i^i)^i = -i$$

And, just for kicks, if we squared that crazy result:

$$((i^i)^i)^2 = -1$$

It's "just" twice the rotation: 2 is a regular number, and it doubles our rotation rate to a full -180 degrees in a unit of time. Or, you can look at it as applying -90 degree rotation twice.

At first blush, these are really strange exponents. But with our analogy we can take them in stride.

11.7 Mixed Growth

We can have both real and imaginary growth: the real portion scales us up, and the imaginary part rotates us around:

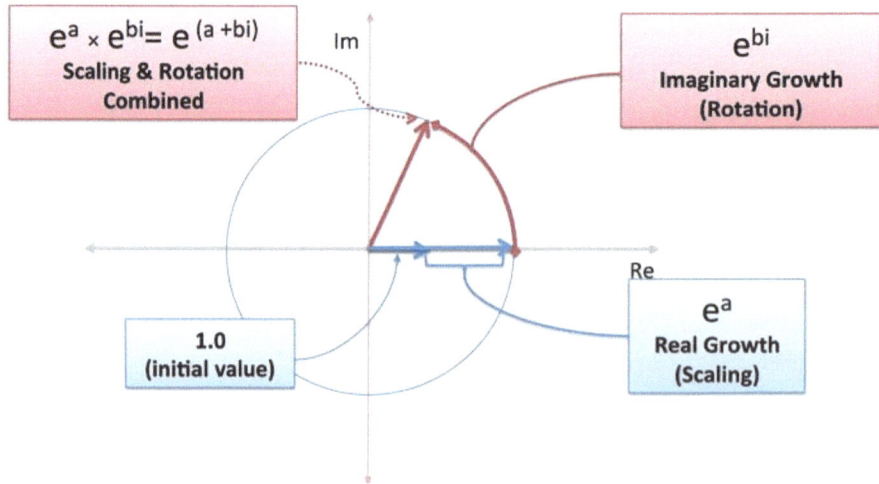

A complex growth rate like (a + bi) is a mix of real and imaginary growth. The real part a, means "grow at 100% for a seconds" and the imaginary part b means "rotate for b seconds". Remember, rotations don't get the benefit of compounding since you keep 'pushing' in a different direction – rotation adds up linearly.

With this in mind, we can represent any point on any sized circle using (a+bi)! The radius is e^a and the angle is determined by e^{ib}. It's like putting the number in the expand-o-tron for two cycles: once to grow it to the right size (a seconds), another time to rotate it to the right angle (b seconds). Or, you could rotate it first and then grow!

Let's say we want to know the growth amount to get to 6 + 8i. This is really asking for the natural log of an imaginary number: how do we grow e to get 6 + 8i?

- Radius: How big of a circle do we need? Well, the magnitude is $\sqrt{6^2 + 8^2} = \sqrt{100} = 10$. Which means we need to grow for $ln(10) = 2.3$ seconds to reach that amount.

- Amount to rotate: What's the angle of that point? We can use arctan to figure it out: $atan(8/6) = 53$ degrees = .93 radian.

- Combine the result: $ln(6 + 8i) = 2.3 + .93i$

11.8 Why Is This Useful?

At a base level, Euler's formula gives us another way to describe motion in a circle. But we could already do that with sine and cosine – what's so special?

It's all about perspective. Sine and cosine describe motion in terms of a *grid*, plotting out horizontal and vertical coordinates.

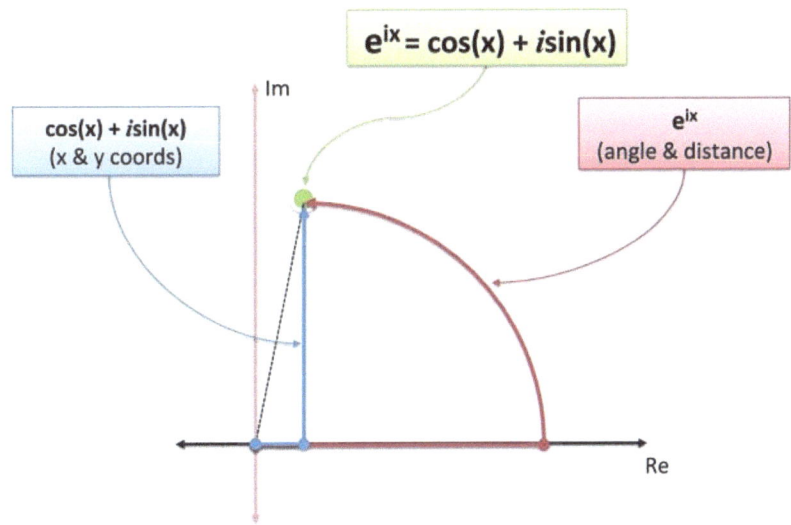

Euler's formula uses polar coordinates – what's your angle and distance? Again, it's two ways to describe motion:

- Grid system: Go 3 units east and 4 units north

- Polar coordinates: Go 5 units at an angle of 71.56 degrees

Depending on the problem, polar or rectangular coordinates are more useful. Euler's formula lets us convert between the two to use the best tool for the job. Also, because e^{ix} can be converted to sine and cosine, we can rewrite every trig formula and identity into variations of e (which is extremely handy – no need to memorize $sin(a + b)$).

But utility, schmutility: it's beautiful that every rotation, every growth rate, and every number (complex or imaginary) is a variation of e. Euler's formula is considered one of the most elegant in all of math – and it really is possible to understand why.

INTRODUCTION TO CALCULUS

I have a love/hate relationship with calculus: it demonstrates the beauty of math and the agony of math education.

Calculus relates topics in an elegant, brain-bending manner. My closest analogy is Darwin's Theory of Evolution: once understood, you start seeing Nature in terms of survival. You understand why drugs lead to resistant germs (survival of the fittest). You know why sugar and fat taste sweet (encourage consumption of high-calorie foods in times of scarcity). It all fits together.

Calculus is similarly enlightening. Don't these formulas seem related in some way?

Circle and Sphere Fun Facts

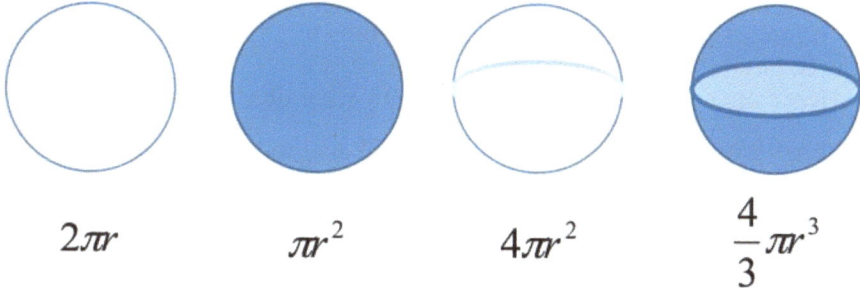

$$2\pi r \qquad \pi r^2 \qquad 4\pi r^2 \qquad \frac{4}{3}\pi r^3$$

They are. But most of us learn these formulas independently. Calculus lets us start with "circumference = $2\pi r$" and figure out the others — the Greeks would have appreciated this.

Unfortunately, calculus can epitomize what's wrong with math education. Most lessons feature contrived examples, arcane proofs, memorization and abstract symbol manipulation that body slam our intuition and enthusiasm before they can put on their gloves.

It really shouldn't be this way.

12.1   Ok Bub, What's Your Great Idea?

Feisty, are we? Well, here's what I won't do: recreate the existing textbooks. If you need answers *right away* for that big test, there's plenty of websites, class videos and 20-minute sprints to help you out.

Instead, let's share the core insights of calculus. Equations aren't enough — I want the "aha!" moments that make everything click.

Formal mathematical language is one just one way to communicate. Diagrams, animations, and just plain talkin' can often provide more insight than a page full of proofs.

12.2   But Calculus Is Hard!

I think anyone can appreciate the core ideas of calculus. We don't need to be writers to enjoy Shakespeare.

It's within your reach if you know algebra and have a general interest in math. Not long ago, reading and writing were the work of trained scribes. Yet today that can be handled by a 10-year old. Why?

Because we expect it. Expectations play a huge part in what's possible. So *expect* that calculus is just another subject. Some people get into the nitty-gritty (the writers/mathematicians). But the rest of us can still admire what's happening, and expand our brain along the way.

It's about how far you want to go. I'd love for everyone to understand the core concepts of calculus and say "whoa".

12.3   So What's Calculus About?

Some define calculus as "the branch of mathematics that deals with limits and the differentiation and integration of functions of one or more variables". It's correct, but not helpful for beginners.

Here's my take: Calculus does to algebra what algebra did to arithmetic.

- **Arithmetic** is about manipulating numbers (addition, multiplication, etc.).

- **Algebra finds patterns between numbers:** $a^2 + b^2 = c^2$ is a famous relationship, describing the sides of a right triangle. Algebra finds entire sets of numbers — if you know a and b, you can find c.

- **Calculus finds patterns between equations:** you can see how one equation ($circumference = 2\pi r$) relates to a similar one ($area = \pi r^2$).

Using calculus, we can ask all sorts of questions:

- How does an equation grow and shrink? Accumulate over time?

- When does it reach its highest/lowest point?

- How do we use variables that are constantly changing? (Heat, motion, populations, ...).

- And much, much more!

Algebra & calculus are a problem-solving duo: calculus finds new equations, and algebra solves them. **Like evolution, calculus expands your understanding of how Nature works.**

12.4 An Example, Please

Let's walk the walk. Suppose we know the equation for circumference ($2\pi r$) and want to find area. What to do?

Realize that a filled-in disc is like a set of Russian dolls.

Dissecting a Circle

Disc ~ Rings

Here are two ways to draw a disc:

- Make a circle and fill it in

- Draw a bunch of rings with a thick marker

The amount of "space" (area) should be the same in each case, right? And how much space does a ring use?

Well, the very largest ring has radius "r" and a circumference $2\pi r$. As the rings get smaller their circumference shrinks, but it keeps the pattern of 2π times current radius. The final ring is more like a pinpoint, with no circumference at all.

Unroll the Rings

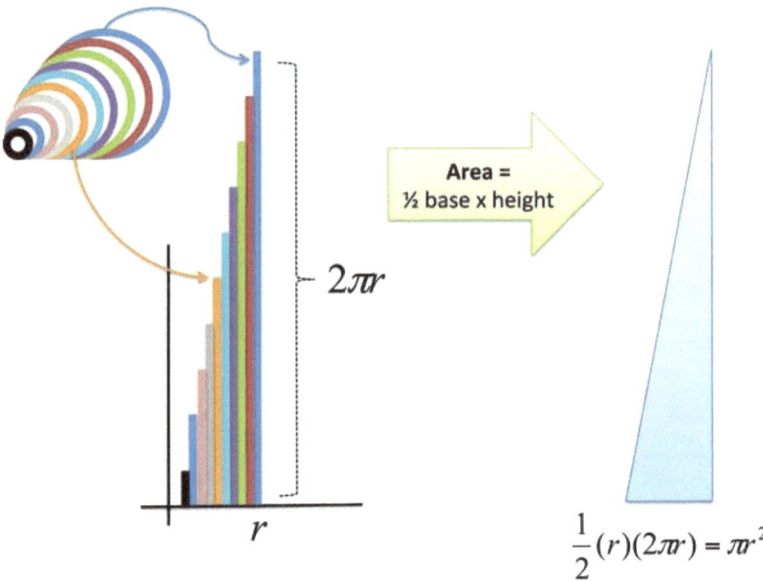

Area =
½ base x height

$2\pi r$

r

$$\frac{1}{2}(r)(2\pi r) = \pi r^2$$

Now here's where things get funky. **Let's unroll those rings and line them up.** What happens?

- We get a bunch of lines, making a jagged triangle. But if we take thinner rings, that triangle becomes less jagged. Taking small segments for more accuracy is a tenet of calculus.

- One side has the smallest ring (0) and the other side has the largest ring ($2\pi r$)

- We have rings going from radius 0 to up to "r". For each possible radius (0 to r), we just place the unrolled ring at that location.

- The total area of the "ring triangle" $= \frac{1}{2} base \cdot height = \frac{1}{2} r(2\pi r) = \pi r^2$, which is the formula for area!

Yowza! The combined area of the rings = the area of the triangle = area of circle!

This was a quick example, but did you catch the key idea? We took a disc, split it up, and put the segments together in a different way. Calculus showed us that a disc and ring are intimately related: a disc is really just a bunch of rings.

This is a recurring theme in calculus: **Big things are made from little things.** And sometimes the little things are easier to work with.

12.5   A Note On Examples

Many calculus examples are based on physics; that's great, but it can be hard to relate. Honestly, how often do you work with *the equation for velocity* for an object? Less than once a week, if that.

I prefer starting with physical, visual examples because it's how our minds work. That ring/circle thing we made? You could build it out of several pipe cleaners, separate them, and straighten them into a crude triangle to see if the math really works. That's just not happening with your velocity equation.

12.6   A Note On Rigor (For the Math Geeks)

I can feel the math pedants firing up their keyboards. Just a few words on "rigor".

Did you know we don't learn calculus the way Newton and Leibniz discovered it? They used intuitive ideas of "fluxions" and "infinitesimals" which were replaced with limits because **"Sure, it works in practice. But does it work in theory?"**.

We've created complex mechanical constructs to "rigorously" prove calculus, but have lost our intuition in the process.

We're looking at the sweetness of sugar from the level of brain-chemistry, instead of recognizing it as Nature's way of saying "This has lots of energy. Eat it."

I don't want to (and can't) teach an analysis course or train researchers. Would it be so bad if everyone understood calculus to the "non-rigorous" level that Newton did? That it changed how they saw the world, as it did for him?

A premature focus on rigor dissuades students and makes math hard to learn. Case in point: e is technically defined by a limit, but the intuition of growth is how it was discovered. The natural log can be seen as an integral, or the time needed to grow. Which explanations help beginners more?

Let's fingerpaint a bit, and get into the chemistry along the way. Happy math.

Afterword

If all went well, insights should be bubbling up about some of the core tenets of math:

- **Imaginary numbers** let us think about numbers in two dimensions

- **e and the natural log** are universal ways to find the impact of and calculate growth rates

- **The Pythagorean theorem** is a general way to measure and compare distances

- **Radians** let us think about rotation from the mover's point of view

- **Growth rates** can be compounded in many ways, applied for different durations, or even used with 0x as a base

- **Euler's formula** lets us travel in a circle using an imaginary (sideways) growth rate

- **Calculus** helps us look at a formula as a whole, or as a collection of smaller parts

Knowledge isn't about acing a quiz: it's about letting ideas become natural and automatic extensions to the way you think. New concepts snap into place because they're based on a solid intuitive foundation, not a fragile memorized one.

Euler's formula is one of the best examples: it's the "jewel" of mathematics and yet its core can be understood with the right foundation in imaginary numbers, radians, and exponents.

If I can leave you with a thought, it's this: Don't be afraid to admit when an idea doesn't click (it happens to me all the time, it's why I write!). There's always a better explanation out there. Happy math.

–Kalid Azad
Seattle, WA
kalid.azad@gmail.com
http://betterexplained.com

95